Statistics for Ornithologists

D1631067

Second Edition

by

Jim Fowler
De Montfort University, Leicester

and

Louis Cohen
Loughborough University of Technology

BTO Guide No 22

© *British Trust for Ornithology, Jim Fowler and Louis Cohen*

Preface

In this extensively revised second edition we re-affirm our aim to introduce ornithologists to the fundamentals of statistics without swamping them with underlying theory, or presuming any prior knowledge of the subject. We have, however, taken advantage of the extensive feedback we have received since the publication of the first edition (together with JF's scrutiny of hundreds of manuscripts during his term as Editor of *Ringing & Migration*) to tailor the Guide even more closely to the needs of ornithologists. Thus, we have dropped the section on the binomial theory (not having encountered a single ornithological example of its use!) and introduced the Kruskal Wallis and G-tests in response to the growing popularity of these techniques. Moreover, we have radically revised the organisation of much of the material, for example, by bringing all the various applications of confidence intervals into a single chapter called *How good are our estimates?*

Since the publication of the first edition some ten years ago there has been a veritable explosion in the use of personal computers and a proliferation of associated statistical packages. However, it is our conviction that the besy way for ornithologists to get a feel for the warp and weft of their data and the scope of relevant statistical techniques to practice with their own counts and measurements using a hand calculator (we still recommend a *CASIO Scientific* calculator for this purpose). We invite readers to begin their introduction to statistics by reworking some of our examples, but don't be put off by minor differences between your results and ours - they arise out of small rounding errors during calculation.

You will soon progess to your own field notebooks and be on the way to acquiring 'statistical literacy'. Then you will be better placed both to use package programs and to avoid the many temptations and pitfalls that they contain.

In addition to those acknowledged in the preface to the first edition, we are grateful to Chris du Feu for scrutinising the manuscript and for his helpful suggestions. We acknowledge that some of the examples used in this guide are based on our book *Practical Statistics for Field Biologists* (pub. John Wiley).

We remain indebted to our wives, Eurgain and Joyce, without whose forbearance we could not possibly have gathered the research data upon which this guide so heavily depends.

Jim Fowler and Louis Cohen.

Foreword

In the Foreword to the first edition of this guide, Raymond O'Connor wrote 'I can look forward to a generation of ringers and ornithologists using *Fowler and Cohen* as their key to a better understanding of their own and other's ornithology'. He was right to do so, for during the intervening years the book has proved a valuable primer not only for ornithologists but also for many other naturalists and field biologists. With the running down of stocks, the authors have taken the opportunity to revise their original text, to take into account the comments and suggestions made by readers.

Ornithologists, like other students of natural history, are faced with the infinite variety of nature: no two Blue Tits are identical in size or shape: no Blue Tit stays the same weight from hour to hour (never mind from season to season); no two woods hold the same number of Blue Tits and, even in the same wood, the number of Blue Tits varies from year to year. Statistics provides us with standard methods for handling that variation, at least in the respect of numerical information - such as the counts and measurements that fill every birdwatchers notebook. The rules have been worked out both through experience and from mathematical principles. They enable us to reduce the mass of data to simpler, more comprehensible forms and to do so in such a way that our conclusions are sound. So, either to interpret his or her own data or to fully understand many of the papers published by others, the ornithologist needs to know these rules.

The need for a clear, simple introduction to statistical methods is particularly great because many ornithologists have had no formal training in the subject: most bird research in Britain and many other countries is carried out by bankers, builders, accountants, actuaries, school teachers, soldiers, doctors, dentists, farmers, foresters, engineers and engine-drivers, rather than by professional ornithologists. It is at them that this book is aimed (though, if more professionals had the thorough grounding in basic ideas that the book provides, competence as data analysts would be much improved!).

The basic ideas in statistics are simple. They are so simple that there is a temptation not to think about them deeply enough, in one's haste to rush on to what seems to be more advanced topics. *Fowler and Cohen* have not given in to that temptation but have covered the basic, simple ideas well in their opening chapters, which are the ones to which readers should pay particular attention. Get those ideas right and the rest follows easily, especially when explained with the clarity that *Fowler and Cohen* bring to bear.

Jeremy Greenwood

Contents

CHAPTER I
INTRODUCTION

1.1 What do we mean by statistics?

Statistics are a familiar and accepted part of the modern world, and already intrude into the life of every serious ornithologist. We have statistics in the form of Observatory reports, annual ringing totals, nest record cards, moult cards, various censuses, to name just a few. It is impossible to imagine life without some form of form of statistical information being readily at hand.

The word **statistics** is used in two senses. It refers to collections of quantitative information, and to methods of handling that sort of data. The annual ringing report, listing the numbers of each species of bird ringed and recovered, is an example of the first sense in which the word is used. Statistics also refers to the drawing of inferences about large groups on the basis of observations made on smaller ones. The calculation of a national "common bird census index" based upon information gathered in a number of representative study sites illustrates the second sense in which the word is used.

Statistics, then, is to do with ways of organising, summarising and describing quantifiable data, and methods of drawing inferences and generalising upon them.

1.2 Why is statistics necessary?

There are two reasons why some knowledge of statistics is an important part of the competence of every ornithologist. First, statistical literacy is necessary if ornithologists are to read and evaluate their journals critically and intelligently. Statements like "the probability that a first-year bird will be found in the North Sea is significantly greater than for an older one, $\chi^2 = 4.1$, df $= 1$, $P<0.05$" enables the reader to decide the justification of the claims made by the particular author.

A second reason why statistical literacy is important is that if ornithologists are to undertake an investigation on their own account and present their results in a form that will be authoritative, then a grasp of statistical principles and methods is essential. Indeed a programme of work should be planned anticipating the statistical methods that are appropriate to the eventual analysis of the data. Attaching some statistical treatment as an afterthought to make the study seem more "respectable" is unlikely to be convincing.

1.3 Statistics in ornithology

"Laboratory" biologists may have high levels of confidence in the precision and accuracy of the measurements they make. To them, catching Storm Petrels at night by means of tape-recordings might appear a hilarious exercise with ludicrously low levels of reliability. "Field" biologists therefore require special sampling techniques and analytical methods if their assertions are to be regarded with credibility. Often data accumulated do not conform to the sort of symmetrical patterns taken for granted in the common statistical techniques; data may be "messy", irregular or asymmetrical. Special treatments may be necessary before they can be properly evaluated.

1.4 The limitations of statistics

Statistics can help an investigator describe data, design experiments, and test hunches about relationships among things or events of personal interest. Statistics is a tool which helps the acceptance or rejection of the hunches within recognised degrees of confidence. They help to answer questions like, *'if my assertion is challenged, can I offer a reasonable defence?'*; or *'Am I justified in spending more time or resources in pursuing my hunch?'*, or *'Can my observations be attributable to chance variation?'*

It should be noted that statistics never prove anything. Rather, they will indicate the likelihood of the results of an investigation being the product of chance.

1.5 The purpose of this guide

The objectives of this guide stem from the points made in Sections 1.2 and 1.3 above. First, the guide aims to provide ornithologists with sufficient grounding in statistical principles and methods to enable them to read and understand the reports and journals they read. Second, the text aims to present ornithologists with a variety of the most appropriate statistical tests for their problems. Third, guidance is offered on ways of presenting the statistical analyses, once completed.

CHAPTER 2
MEASUREMENT AND SAMPLING CONCEPTS

2.1 Populations, samples and sampling units

Ornithologists will be familiar with the term "population" in an ecological sense. A population comprises all the individuals of a species that interact sexually with each other to maintain a homogeneous gene pool. In statistics, the term **population** is extended to mean any collection of individual items or **units** which are the subject of investigation. Characteristics of a population which differ from individual to individual are called **variables**. Length, mass, age, feeding rate, moult score, arrival date, proximity to a neighbour, number of parasites, clutch size, to name but a few, are examples of biological variables to which numbers or values can be assigned. Once numbers or values have been assigned to the variables they can be measured.

Because it is rarely practicable to obtain measures of a particular variable from all items or units in a population, the investigator has to collect information from a smaller group or sub-set which represents the group as a whole. This sub-set is called a **sample**. Each unit in the sample provides a record, such as a measurement, which is called an **observation**. The relationship between the terms we have introduced in this section is summarised below:

Observation:	132 mm
Variable:	wing length
Sample unit:	a starling from a communal roost
Sample:	those starlings which are captured in the roost and are measured
Statistical population:	all starlings in the roost which are available for capture and measurement
Biological population:	the biological population may well include birds that are not available for capture (e.g. mates that are roosting elsewhere) and are therefore not part of the statistical population. Alternatively, if the roost comprises a mixture of resident birds and winter immigrants, the statistical population might include components of more than one biological population.

A collection of observations (perhaps in more than one sample) is called **data**. The word data is always plural: thus, "these data were recorded last year".

2.2 Counting things – the sampling unit

Ornithologists often count the number of objects in a group or collection. If the number is to be meaningful, the dimensions of the collection have to be specified. A collection with specified dimensions is called a **sampling unit**; a set of sampling units comprises a sample. An observation is, of course, the number of objects in a particular sampling unit. Examples of sampling units are:

Observation	Sampling unit
The number of birds counted or caught	Unit area; unit length of beach or cliff ledge; unit time; unit length of mist net
Number of parasites carried	A single bird
Clutch size	A single clutch

When observations are counts, the statistical population has nothing to do with the objects we are counting, even when they are organisms. The following example illustrates the point:

Observation: 5

Variable: number of Puffin burrows

Sampling unit: a square ("quadrat") of stated dimensions

Sample: the number of quadrats (sampling units) examined

Statistical population: the number of quadrats it is *possible* to mark out in the whole of the study area. The potential number of units in the population depends on the chosen dimensions of the sampling unit.

The main difference between 'measuring' and 'counting' is that we have no control over the dimensions of a unit in a sample when we are measuring; when counting we are able to choose the dimensions of the sampling unit. Remember that the contents of a trap, net or quadrat is a sample if we are measuring them, but only a unit in a sample if we are counting them.

It is always worthwhile to ask the question *'from which population are my sampling units drawn?'* The answer may not always be as obvious as in the example of the puffin burrows. If an hour's catch in a mist-net is a sampling unit, then five such units comprise a sample – but from which population are these sampling units drawn? It is regarded as being the total number of nets that could be set, covering the whole of the study area. In that case there would be no birds left to catch, and so statistical populations are sometimes notional or **hypothetical**. Samples drawn from hypothetical populations are more often used for comparative purposes, e.g. to compare one year with another in the Constant Effort Scheme.

2.3 Random sampling

We say in Section 2.1 that a sample *represents* the population from which it is drawn. If the sample is to be truly representative, the units in the sample must be drawn **randomly** from the population; that is to say, in a manner that is free from **bias**. In other words, each unit in the sample must have an equal chance of being drawn.

As an example of a possible source of bias, consider an ornithologist who wishes to measure the average weight of Blue Tits visiting gardens. In order to boost the catch, birds are attracted to the garden with food. It is plausible that hungry birds are less shy than well-fed ones, and this might increase the chance of them being drawn from the population. If hungry birds are lighter than well-fed ones, our ornithologist's sample may not be a fair representation of the whole population. In practice, perfectly random sampling may be very difficult indeed. For example, siting a mist net randomly would involve drawing a map of the study site with a grid superimposed. The mist net would be sited at the intersection of pairs of coordinates prescribed by a random number generating computer, or tables. A few examples of possible sources of bias are listed below:

Activity	Source of bias
Setting mist net high	Bias against Wrens and Grasshopper Warblers
Setting mist net low	Bias against canopy feeding birds
Catching in moult period	Bias against skulking birds in moult
Attracting birds with bait	Bias for hungry (underweight) birds
Attracting birds with tape lure	Bias for one sex or age class

Statistical analysis is frequently carried out on the assumption that samples are random. If, for any reason, that assumption is false and bias is present in the sampling procedure, then the information gained from the sample may not be properly extrapolated to the population. Unfortunately, it is rarely possible do more than guess how great bias may be. This severely reduces the confidence which can be placed in estimations based on sampling data. Since most sources of bias arise from the methodology adopted, procedures should always be very fully described. When a source of bias is suspected it should be acknowledged and taken into account in the interpretation of the results:

> "... gravid females were not deparasitised in the nesting season in order to avoid the risk of damaging unlaid eggs, and it is accepted that this precaution could introduce bias into the results" (*Bird Study* 30: 240).

A particularly insidious source of bias, observer bias, conscious or subconscious, is notoriously difficult to avoid when gathering data in support of a particular hunch!

The practical aspects of obtaining random samples is a large area in itself, partly because the techniques used by ornithologists (and biologists in general) are so diverse. We suggest you consult Southwood (1978) as a standard work on this subject (see Bibliography). See *Ringing & Migration* 1: 105 for an interesting discussion of sample bias.

2.4 Independence

Many statistical methods assume that observations in a sample are **independent**. That is to say, the value of any one observation in a sample is not inherently linked to that of another. A problem would arise, for example, if repeated weight measurements of a recaptured bird were included in a sample. If the bird happened to be exceptionally large (or small) the several obtained measurements, though different in value, are *not* independent and would therefore distort the sample.

2.5 Statistics and parameters

The measures which describe a variable of a sample are called statistics. It is from the sample statistics that the **parameters** of a population are estimated. Thus, the average wing length of a sample of birds is the statistic which is used to estimate the average wing length parameter of the population. The average number of puffin burrows in a random sample of quadrats estimates the average number of burrows per quadrat in the whole population of quadrats.

In estimating a population parameter from a sample statistic, the number of units (observations) in a sample can be critical. Some statistical methods depend on a minimum number of sampling units and, where this is the case, it should be borne in mind before commencing fieldwork. Whilst it is true that larger samples will invariably result in greater statistical confidence, there is nevertheless a 'diminishing returns' effect. In many cases the time, effort and expense involved in collecting very large samples might be better spent in extending the study in other directions. We offer guidance as to what constitutes a suitable sample size for each statistical test as it is described.

2.6 Descriptive and inferential statistics

Descriptive statistics are used to organise, summarise and describe measures of a sample. No predictions or inferences are made regarding population parameters. **Inferential** (or **deductive**) statistics, on the other hand, are used to infer or predict population parameters from sample measures. This is done by a process of deductive reasoning based on the mathematical theory of **probability**. Fortunately, only a very minimal knowledge of the mathematical theory of probability is needed in order to apply the rules of the statistical methods, and the little that is needed will be explained. However, no one can predict exactly a population parameter from a sample statistic, but only indicate with a stated degree of confidence within what range it lies. The degree of confidence depends on the sample selection procedures and the statistical techniques used.

2.7 Parametric and non-parametric statistics

Statistical methods commonly used by ornithologists fall into one of two classes – **parametric** and **non-parametric**. Parametric methods are the oldest, and although most often used by statisticians, may not always be the most appropriate for analysing biological data. Parametric methods make strict assumptions which may not always hold true, for example, that the data are "normally" distributed (see Chapter 7).

More recently, non-parametric methods have been devised which are not based upon stringent assumptions. These are frequently more suitable for processing biological data. Moreover they are generally simpler to use since they avoid the laborious and repetitive calculations involved in some of the parametric methods. The circumstances under which a particular method should be used will be described as it arises. A summary showing which methods should be applied in particular circumstances is provided in Section 10.8.

CHAPTER 3
PROCESSING DATA

3.1 Scales of measurement

Variables measured by ornithologists can be either **discontinuous** or **continuous**. Values of discontinuous variables assume integral, or whole numbers and are usually counts of things (frequencies). On the other hand, values of continuous variables may, in principle, fall at any point along an uninterrupted scale, and are usually measurements (length, weight, temperature, etc.). Measurement values may sometimes appear to be integral whole numbers if the recorder elects to measure to the nearest whole unit; this does not, however, obviate the fact that there can be intermediate values. The distinction between 'count data' and 'measurement data' is an important one which will be referred to frequently.

Generally, four levels of measurement are recognised. They are referred to as **nominal**, **ordinal**, **interval**, and **ratio** scales. Each level has its own rules and restrictions; moreover, each level is hierarchical in that it incorporates the properties of the scale below it.

3.2 The nominal scale

The most elementary scale of measurement is one which does no more than identify categories into which individuals may be classified. The categories have to be mutually exclusive, i.e. it should not be possible to place an individual in more than one category. The nominal level of measurement is often used by ornithologists. For example, species, sex, age class, colour and habitat type are all nominal categories into which count data can be assigned.

The 'name' of a category can of course be substituted by a number – but it will be a mere label and have no numerical meaning. Thus, if Blue Tits are coded 1, Coal Tits 2, Great Tits 3, Willow Tits 4 and Marsh Tits 5 they can then be listed, 1, 2, 3, 4, 5 but the sequence has no more mathematical significance than if they had been listed 4, 2, 1, 5, 3. They are still nominal categories.

3.3 The ordinal scale

The ordinal scale incorporates the classifying and labelling function of the nominal scale, but in addition brings to it a sense of order. Ordinal numbers are used to indicate **rank order**, but nothing more. The ordinal scale is used to arrange (or rank) individuals into a sequence ranging from the highest to the lowest, according to the variable being measured. Ordinal numbers assigned to such a sequence may not indicate absolute quantities, nor can it be assumed that the intervals between adjacent numbers on the scale are equal. In ornithological studies, the crude 'abundance scales' used in atlassing work are a good example of ordinal scales:

Abundance score	Description	
5	Abundant	– over 1000 individuals
4	Frequent	– 100 – 1000 individuals
3	Common	– 20 – 100 individuals
2	Uncommon	– 5 – 20 individuals
1	Rare	– 1 – 5 individuals

In this scale there is no simple relationship between the numerical values on the abundance scale. 'Frequent' does not mean twice 'uncommon', but it will always be ranked above 'uncommon'.

3.4 The interval scale

As the term 'interval' implies, in addition to rank ordering data, the interval scale allows the recognition of *precisely how far apart* are the units on the scale. Interval scales permit certain mathematical procedures untenable at the nominal and ordinal levels of measurement. Because it can be concluded that the difference between the values of, say, the 8th and 9th points on the scale is the same as that between the 2nd and 3rd, it follows that the intervals can be added or subtracted but because a characteristic of interval scales is that they have *no absolute zero point* it is not possible to say that the 9th value is three times that of the third. To illustrate this, date is a very widely used interval scale. If the first-arrival dates of four species of warbler are, respectively, the 1st, 5th, 10th, and 15th May, the interval between each point on the scale (1 day) is equal and the fourth species took 10 days longer to arrive than the second. It did not take three *times* as long, however, any more than it took 15 times longer to arrive than the first species! Another interval scale is temperature: 10°C is not twice as hot as 5°C because the zero on the scale in question (Celsius) is not absolute.

3.5 The ratio scale

The highest level of measurement, which incorporates the properties of the interval, ordinal and nominal levels, is the ratio scale. A ratio scale includes an absolute zero, it gives a rank ordering and it can be used for simply labelling purposes. Because there is an absolute zero, all of the mathematical procedures of addition, subtraction, multiplication and division are possible. Measurements of length and weight fall on ratio scales. Thus, a length of 150 mm *is* three times as long as one of 50 mm.

In practice, the mathematical properties of interval and ratio scales are similar and as no statistical test in this Guide will distinguish between them, we shall refer to them both as 'interval' scales.

3.6 Conversion of interval observations to an ordinal scale

Usually, observations made on interval scales allow the execution of more sensitive statistical analyses. Sometimes, however, interval data are not suitable for certain methods. Perhaps, because there are too few observations, we are forced to downgrade them to an ordinal rank scale for use in non-parametric methods. The following measurements (mm) are ranked in increasing size in the top line. Their rank (ordinal) scores are underneath:

Length:	31.0	31.4	32.3	33.1	33.5	34.9	35.0	36.6	37.2	38.8
Rank score:	1	2	3	4	5	6	7	8	8	10

If large numbers of observations of a variable are collected, it is almost inevitable that some of the observations will be equal in value. Their ranks will also be tied and these have to be dealt with correctly. Since some statistical tests which we describe later depend on the ranking of observations, we take the opportunity now of dealing with the problem of ranking **tied observations**.

Where tied observations occur, each of them is assigned the average of the ranks that would have been given if there were no ties. To illustrate this, a set of measurements rounded to the nearest whole number is given below. For convenience they are presented in ascending order and adjacent tied scores are underlined:

25 26 <u>27 27</u> 28 29 <u>30 30 30</u> 31 32 <u>33 33 33 33</u> 34 35 <u>36 36 36 36 36</u> 37

If we try to rank these, the single extreme values of 25 and 37 will clearly be ranked 1 and 23, respectively. The two values of 27 together occupy the ranks of 3 and 4; they are each assigned the average rank of 3.5. The three values of 30 occupy the ranks 7, 8 and 9. They have an average rank of 8. In similar manner, the four values of 33 are each assigned the rank of 13.5 and the five of 36 the rank of 20. The set of data is rewritten below, with the correct ranks assigned:

Observation:	25	26	27	27	28	29	30	30	30	31	32	33
Rank:	1	2	3.5	3.5	5	6	8	8	8	10	11	13.5

Observation:	33	33	33	34	35	36	36	36	36	36	37
Rank:	13.5	13.5	13.5	16	17	20	20	20	20	20	23

3.7 Derived variables

Sometimes observations are processed in order to generate a **derived number**. Examples of derived variables are *ratios*, *proportions*, *percentages*, *rates*, and various *indices*.

A **ratio** is the simple relationship between two numbers. A sample which contains 18 males and 25 females has a male:female ratio of $18:25$ or $1: \frac{25}{18}$, that is, $1:1.39$. A ratio may be expressed as a fraction. In the last example, the ratio of males to females is $18/25 = 1/1.39$. When the fraction is reduced to a decimal number, it is often called a *coefficient*; thus, $1/1.39 = 0.72$.

A **proportion** is the ratio of a part to a whole. Pied Wagtails in winter spend about 7.65 hours each day foraging (*British Birds* 75:261). The proportion of the day occupied by this activity is $7.65:24 = 0.32$. If a proportion is based on counts of things it may be referred to as a *proportional frequency*, that is, the ratio of the numbers of individuals in a particular category to the total number in all categories.

Example 3.1
The number of five species of tit counted in a habitat are: Blue Tit 90; Coal Tit 31; Great Tit 72; Willow Tit 19; Marsh Tit 17. What is the proportion of each species in the sample?

The total number N of tits is $90 + 31 + 72 + 19 + 17 = 229$. The proportion is given by:

$$p_i = \frac{n_i}{N}$$

where p_i is the proportion of a particular category, n_i is the number of individuals in a particular category and N is the total number in all categories. Therefore:

Species	n_i	n_i/N	p_i	Percentage
Blue Tit	90	90/229	0.39	39
Coal Tit	31	31/229	0.14	14
Great Tit	72	72/229	0.31	31
Willow Tit	19	19/229	0.08	8
Marsh Tit	17	17/229	0.07	7
	$N = 229$		$\Sigma p_i = 1$	

Notice that the sum of the individual proportions equals 1 (subject to rounding errors). Σ means "the sum of".

When a proportion is multiplied by 100 it is called a **percentage**. The percentage of each species of tit in the example described in Example 3.1 is included in the table above.

A **rate** is the ratio of an observation to a period of time. Rates are useful for expressing such variables as growth, population change, and movement. If a pigeon flies 1728 km in 24 h, the distance:time ratio is 1728:24 = 72:1. The rate is 72 km/h.

Note that the word "rate" often becomes synonymous with "proportion": thus, **survival rate** is the proportion of individuals alive in one year which survive to the next.

Statistical techniques may be performed upon derived variables; sometimes the data first have to be converted or **transformed** (see, for example, Section 8.1).

3.8 Logarithms

Logarithms are a special form of derived variable frequently used by scientists of all kinds. The logarithm of a number is obtained simply by entering the number in a scientific calculator and pressing the "log" key. An understanding of the theory of logarithms is by no means essential in the use of statistics, but a brief explanation may be helpful.

A logarithm is the *power* to which a *base* must be raised to produce a given number. The base can be any chosen number. For example, if the base is 2, the given number 8, then the power is 3:

$$8 = 2 \times 2 \times 2 = 2^3$$

That is to say, the logarithm of 8 to the base 2 is 3, or $\log_2(8) = 3$.

The most widely used logarithmic base is 10. In a similar way:

$$1000 = 10^3, \qquad \text{therefore} \qquad \log_{10}(1000) = 3$$
$$100 = 10^2, \qquad \text{therefore} \qquad \log_{10}(100) = 2$$
$$10 = 10^1, \qquad \text{therefore} \qquad \log_{10}(10) = 1$$

It follows that the $\log_{10}$ of a number between 10 and 100 must be between 1 and 2, namely "one point something". Fortunately we do not have to calculate these intermediate values because a calculator (or tables) will give the answer. Because logarithms to the base 10 are in such wide use, the subscript 10 is often omitted and log(15) means $\log_{10} 15 = 1.176$. The logarithm of 1 is always zero and the logarithm of 0 is always minus infinity, no matter what the base. One advantage of using logarithms is that they compress values spanning several orders of magnitude onto a convenient scale. Thus, log(10) = 1; log(1,000,000) = 6. See section 8.2 for a practical example of how logarithms can "squash up" a scale.

Having obtained a logarithm, it may be necessary to convert it back to the original number. This is called taking the antilogarithm (or antilog). On a calculator this is obtained by entering the number for which the antilog is required, and pressing the *INV* key before the log key. Thus, antilog (1.43) = 26.92.

There is another base upon which biologists frequently express logarithms and that is the value of a natural constant e. The value of e is 2.718 and although this may seem a rather strange number to use as a logarithmic base, logarithms to the base e are suitable for more advanced mathematical treatment. Logarithms to the base e are called **natural logarithms** are denoted by $\log_e$ or, simply, *ln*. Thus:

$$\log_e(25) = ln(25) = 3.219$$

which is the same as writing $25 = 2.718^{3.219}$. To obtain a natural logarithm on a calculator, enter the number and press the key marked *ln*.

3.9 The precision of observations

When an observation is of a discrete variable, that is a **count**, we are usually sure of its precision. A nest may have *exactly* four eggs in it. A **measurement**, on the other hand, is never exact; it is precise only to within certain limits.

If the diameter of a pond is measured with a tape marked in 1-metre intervals, measurements are precise to the nearest whole metre. An observation of 10 m can be recorded as 10 ± 0.5 m. This implies that all distances between the limits of 9.5 m and 10.5 m are recorded as 10 m, as shown in Fig. 3.1.

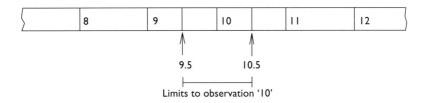

Fig. 3.1 The limits of an observation

We could use a more finely graduated scale, for example a tape marked in 10 cm (0.1 m) intervals. Each observation is then precise to the nearest 0.1 m and an observation of 10.6 m is written as 10.6 ± 0.05 m; that is, all distances between 10.55 m and 10.65 m are recorded as 10.6 m.

We could continue increasing the precision of measurements by using a metre rule graduated in millimetres, Vernier callipers capable of recording to 0.01 mm, or a microscope eye-piece graticule down to 0.001 mm. In each case, the measurement is precise to *plus or minus half the interval spanned by the last measured digit*; in the case of the graticule measurement this is $\pm$ 0.0005 mm. An observation of 1.364 mm is within an interval spanning 1.3635 mm to 1.3645 mm.

Note the distinction between *precision* and *accuracy*. An expensive spring balance might be precise, weighing to 0.1 g, but if it is badly adjusted it will not be accurate. A broken clock is accurate twice a day!

3.10 The frequency table

Collected data should be organised and summarised in a form that allows further interpretation and analysis. In Table 3.1 are the wing length measurements (to the nearest whole millimetre) of 100 Robins.

Table 3.1 Wing lengths of 100 Robins

76	73	75	73	74	74	74	74	74	77
74	72	75	76	73	71	73	80	75	75
68	72	78	74	75	74	69	77	77	72
72	76	76	77	70	77	72	74	77	76
78	72	70	74	76	72	73	71	74	74
75	79	75	74	75	74	71	73	75	73
75	70	73	75	70	72	72	71	76	73
74	76	74	75	74	76	75	75	73	73
78	74	73	75	74	73	72	76	73	76
74	71	72	71	79	78	69	77	73	71

The measurements are presented in the order in which the observations were recorded and are therefore *ungrouped*. A quick scan of Table 3.1 reveals that particular values are repeated a number of times: there are, for example, five values of 74 mm in the top row alone. The value of *74 mm* is called a **frequency class** and, rather than record all values individually (as in the table), it is more economical of space and more revealing to *group* the data into all the frequency classes. We should remember that each frequency class is a **class interval** with implied limits of ±0.5 mm. Thus, all lengths between 73.5 mm and 74.5 mm are placed in the '74 mm' class. The grouped observations are shown in Table 3.2.

Table 3.2 Grouped lengths of 100 Robin wing lengths

Implied class interval	Frequency class x (mm)	Tallies	Frequency f
67.5 – 68.5	68	1	1
68.5 – 69.5	69	11	2
69.5 – 70.5	70	1111	4
70.5 – 71.5	71	1111111	7
71.5 – 72.5	72	1111111111	11
72.5 – 73.5	73	111111111111111	15
73.5 – 74.5	74	11111111111111111111	20
74.5 – 75.5	75	111111111111111	15
75.5 – 76.5	76	1111111111	11
76.5 – 77.5	77	1111111	7
77.5 – 78.5	78	1111	4
78.5 – 79.5	79	11	2
79.5 – 80.5	80	1	1
			$n = \Sigma f = 100$

The data in table 3.2 are grouped into columns: implied class interval, frequency class x and frequency f. The tallies are presented in this example to give a visual appreciation of how the frequencies are distributed between the frequency classes. The two columns x and f represent a **frequency table**. Although this particular table has been constructed for interval measurements, frequency tables can also be constructed for count data and for nominal and ordinal scales. The manner in which frequencies are distributed between the frequency classes is described as a **frequency distribution**.

3.11 Bivariate data

Ornithologists frequently obtain more than one observation from a unit in a sample. A bird may be weighed and measured; a quadrat may provide a count of puffin burrows and a soil depth; a ringing recovery may provide a distance and a time.

A set of observations of *two* variables from each item or unit in a sample is called bivariate data. There are statistical methods available for analysing **bivariate data** (see Chapters 12 & 13).

CHAPTER 4
PRESENTING DATA

4.1 Introduction
One drawback in presenting data in the form of a frequency distribution table is that the information contained there does not become immediately apparent unless the table is studied in detail. To simplify the interpretation of the information, and to pin-point patterns and trends, the data are often processed further and transformed into a visual presentation. The most common methods of presenting data are based upon graphical techniques. In this chapter we describe methods suitable for presenting ornithological data.

4.2 Dot diagram or line plot
The dot diagram is a method of presenting data which gives a rough but rapid visual appreciation of the way in which the data are distributed. It consists of a horizontal line marked out with divisions of the scale on which the variable is being measured. A dot representing each observation is placed at the appropriate point on the scale. If certain observations are repeated, the dots are simply stacked on top of each other. Fig. 4.1 shows two dot diagrams, each involving 16 sampling units:

(a) Sixteen Blue Tit wing lengths: observations are scattered about 63 mm.
(b) Fleas on 16 Blackbirds: 7 Blackbirds don't have any.

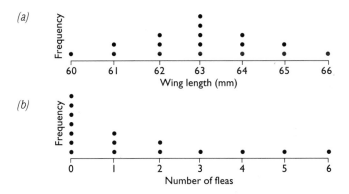

Fig 4.1 Dot diagrams showing frequency distributions of (a) 16 Blue Tit wing lengths; (b) fleas on 16 Blackbirds.

Although dot diagrams are invaluable for the preliminary analysis of data, they are seldom used as the form of final presentation.

4.3 Bar graph
Portraying information by means of a bar graph is particularly useful when dealing with data gathered from discrete variables that are measured on a nominal scale. A bar graph uses vertical lines (i.e. bars) to represent discrete categories of data, the length of the line being proportional to the frequencies within that category.

Suppose 31 nest boxes are placed in a wood, and 15 become occupied by Blue Tits, 10 by Great Tits, 4 by Tree Sparrows and 2 by Nuthatches. A frequency table may be constructed:

	f
Blue Tit	15
Great Tit	10
Tree Sparrow	4
Nuthatch	2
	$n = 31$

Using these data, a bar graph may be constructed, as shown in Fig. 4.2.

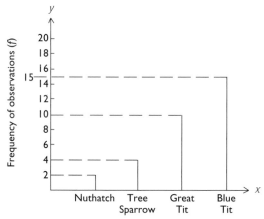

Fig. 4.2 Bar graph showing nest box occupancy

In its final form, the horizontal dashed lines are omitted; they are included here to show that the height of the bar corresponds to the respective frequency.

When observations are counts of things the bar graph is a useful way to present a frequency distribution. Illustrators often replace each bar with a vertical rectangle, or block, whose adjacent sides are touching. The frequency distribution of flea counts shown as a dot diagram in Fig. 4.1 is shown as a bar graph in Fig. 4.3, where the height of each block is still proportional to the frequency in each category because the width of each block is equal. When presented in this form the diagram is usually referred to as a **histogram**. Histograms are especially useful for presenting frequency distributions of observations measured on continuous variables, as we show in the next section.

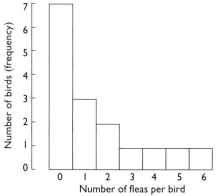

Fig. 4.3 Frequency distribution of flea counts

4.4 Histogram

The histogram is especially useful for presenting distributions of observations of **continuous** variables. In a histogram the area of each block is proportional to the frequency. The area of a single histogram block is found by multiplying the width of the block (the class interval) by the height (frequency).

Example 4.1

In some areas a spiny fish called the sea scorpion forms a high proportion of the diet of the Cormorant (*Seabird* **14**:21–26). 150 sea scorpions were recovered from regurgitates of Cormorant chicks during ringing operations and measured to the nearest whole millimetre. Measurements obtained are expressed in the form of a frequency table:

Length of fish (mm)	Number of fish (frequency)
100 – 109	7
110 – 119	16
120 – 129	19
130 – 139	31
140 – 149	41
150 – 159	23
160 – 169	10
170 – 179	3

The frequency distribution is presented in a histogram in Fig. 4.4.

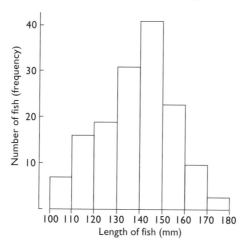

Fig. 4.4 Frequency distribution of sea scorpion lengths

Because the width of each block (class interval) is the same, 10 mm, the height of each block is proportional to the frequency. In Fig 4.4 the area of the first block is 10 x 7 = 70 units, compared with the fifth block which is 10 x 41 = 410 units. The area represented by the last three blocks is (10 x 23) + (10 x 10) + (10 x 3) = 360 units.

Sometimes it is convenient to *aggregate* frequency classes, especially those in the 'tails' of the distribution. The distribution of Example 4.1 is re-written in the table below with the last three classes aggregated:

Length of fish (mm)	Number of fish (frequency)
100 – 109	7
110 – 119	16
120 – 129	19
130 – 139	31
140 – 149	41
150 – 179	36

In the last category, the 36 fish are aggregated into a class spanning 30 mm. The *area* of the block which will represent them is still 360 units, as above. However, since the width of the block is 30 units, the height of the block is 360/30 = 12 units. This is, of course, the average height of the three blocks before they were aggregated. The resulting histogram is shown in Fig. 4.5.

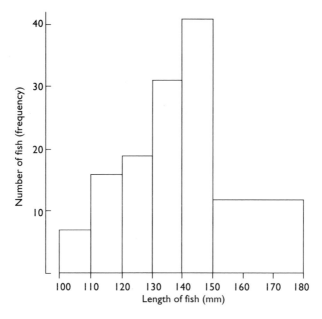

Fig. 4.5 Frequency distribution of fish lengths with aggregated tail

4.5 Frequency polygon and frequency curve

If the mid-point of the top of each block in a histogram is joined by a straight line, a **frequency polygon** is produced. Fig. 4.4 is reproduced with a frequency polygon superimposed in Fig. 4.6. When the number of observations of a continuous variable is large and the unit increments are small, the 'steps' in the histogram tend towards a smooth, continuous curve, called a **frequency curve**. A frequency curve is superimposed (in Fig. 4.7) upon the distribution of 100 Robin wing lengths given in Section 3.10.

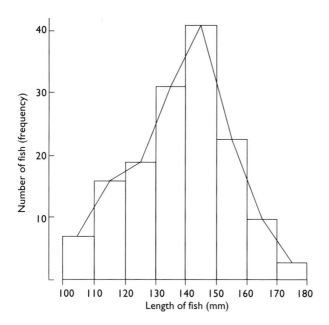

Fig. 4.6 Frequency polygon of sea scorpion lengths

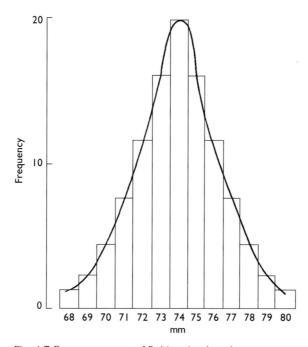

Fig. 4.7 Frequency curve of Robin wing length measurements

4.6 Kite graph

Changes in frequency distribution with altitude are often displayed in the form of a kite graph. In kite graphs, altitude is recorded on the vertical axis, and frequency on the horizontal axis. The frequency data are effectively frequency polygons duplicated as a 'mirror image' on either side of the zero axis to present a visually pleasing symmetrical diagram, sometimes resembling the shape of a kite, which can be helpful in pin-pointing "zones" of distribution. Fig. 4 shows an example of the altitudinal distribution of Stonechats and Whinchats. Note that each percentage is plotted *twice*; thus, 30% is not split into 15% to the left and 15% to the right, but 30% on the side of each axis.

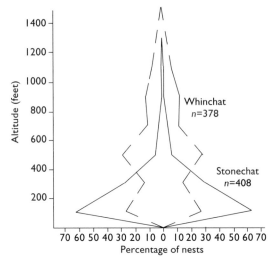

Fig. 4.8 Altitudinal distribution of Stonechat and Whinchat nests recorded at 100 intervals (Bird Study: 24:217).

4.7 Scattergram (Scatter diagram)

When *pairs* of observations of two variables are obtained from each unit in a sample (that is, the data are **bivariate**), a scattergram is used to display the data. In Fig. 4.9 each point on the scattergram represents a single Dunlin for which bill length and weight are indicated. In two dimensional presentations like this it is conventional to refer to the vertical scale as the *y axis* and the horizontal scale as the *x axis*.

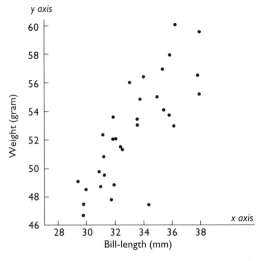

Fig. 4.9 Relationship between bill length and weight in Dunlins (Bird Study 30: 159).

4.8 Circle or pie graph

The pie graph is best suited for displaying data which are *proportions* or *percentages*. If the area of a circle is regarded as 100% it can be divided into sectors (the slices of the pie) which correspond in size to each individual percentage or proportion making up the total. To work out the angle of each sector at the centre of the pie divide each percentage by 100 and multiply by 360, the number of degrees in a circle. Proportions are simply multiplied by 360. It is conventional to sequence the slices in order of size, except for aggregated class ("others") which fill up the space between the smallest and largest. Fig. 4.10 shows bird community structure in two habitats displayed as pie graphs.

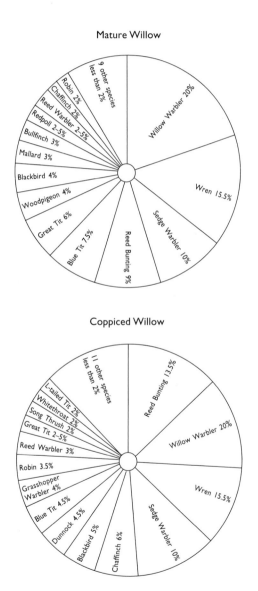

Fig. 4.10 Bird community structure in willow scrub at Leighton Moss (Bird Study 25: 240)

4.9 Map circles

Map circles are used to indicate geographical distribution of observations; variation in circle size (with a reference scale) indicates magnitude. Note, however, that the "size" of a circle may be deceptive because its area increases with the square of its radius. They are therefore best regarded as a visual display of the ordinal level "abundance scale" described in Section 3.3. Many ornithologists will be familiar with their use in the BTO/IWC Atlas of Breeding Birds in Britain and Ireland.

Fig. 4.11 uses map circles to pin-point the location and relative magnitude of inland Arctic Tern

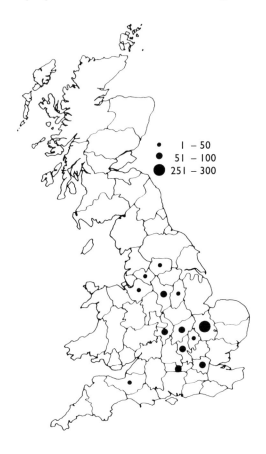

records.

Fig. 4.11 Distribution by counties of inland records of Arctic Terns during 24th–30th April 1981
(British Birds 75: 564).

4.10 Choice of method

The choice of method used in presenting data in a graphical form must rest upon the nature of the original data and the amount of detailed visual information required. Remember, pictorial methods of presentation add nothing to the data that wasn't already there to begin with! Their task is to display it more effectively. A useful booklet *On the graphical presentation of quantitative data* by J.H. Crothers is available from the Field Studies Council, Preston Montford, Shrewsbury S74 1HW.

CHAPTER 5
MEASURING THE AVERAGE

5.1 What is an average?

One meaning of statistics is to do with describing and summarising quantitative data. Any description of a sample of observations must include an aspect which relates to **central tendency**. That is to say, we need to identify a single number close to the centre of the distribution of observations which represents them all. We call this number the **average**; it is often referred to as a measure of **location** because it indicates, on what might be a scale of infinite magnitude, just where a cluster of observations is located. The average is described by one of three commonly used statistics: the **mean**, the **median** and the **mode**. Each has its own application.

5.2 The mean

The mean or, more precisely, the **arithmetic mean** is the most familiar and useful measure of the average. It is calculated by dividing the sum of a set of observations by the number of observations. If it possible to obtain an observation from every single item or sampling unit in the population (for example the weight of every single living Californian Condor, or the numbers of Puffin burrows in every single quadrat that it is possible to mark out on an island) then the mean is symbolised by μ (mu) and is called the **population mean**. More usually, we have to be content with observations from a sample, in which case the **sample mean** is symbolised by $\bar{x}$ ('x-bar'). The sample mean is a direct estimate of the population mean (thus $\bar{x}$ estimates μ). In Chapter 9 we explain how good an estimate it is likely to be. The formulae for calculating the mean are:

$$\mu = \frac{\Sigma x}{N} \qquad\qquad \bar{x} = \frac{\Sigma x}{n}$$

where x is each observation, N is the number of items (observations) in a population, n is the number of observations in a sample and Σ is 'the sum of'.

Example 5.1

Calculate the mean weight of the five birds recorded below:

    8.5 g   9.2 g   7.3 g   6.8 g   10.1 g

1. Obtain the sum of the observations (Σx):

$\Sigma x = 8.5 + 9.2 + 7.3 + 6.8 + 10.1 = 41.9$ g

2. Divide Σx by n, the number of observations:

$$\bar{x} = \frac{\Sigma x}{n} = \frac{41.9}{5} = 8.38 \text{ g}$$

Conventionally, the mean is recorded to one more decimal place than the original data if n is less than 100, to two decimal places if n is between 100 and 999, and so on.

Example 5.2
The number of eggs recorded in five Blue Tit clutches are:

| 8 | 9 | 7 | 11 | 9 |

Calculate the mean number of eggs per clutch.

$$\bar{x} = \frac{\Sigma x}{n} = \frac{44}{5} = 8.8$$

In this example, observations relate to a discrete (non-continuous) variable – a whole number of eggs; the sampling unit is a clutch. Notice, however, that the *mean is a fraction*; it can assume any degree of precision. Therefore a sample mean is a *continuous variable*, even if the observations in the sample are not. The consequences of this important point are considered in Section 9.2.

Example 5.3
The numbers of Gannets seen passing a headland during seven consecutive days are:

| 25 | 4 | 12 | 9 | 202 | 15 | 8 |

Calculate the mean number of Gannets seen per day.

$$\bar{x} = \frac{\Sigma x}{n} = \frac{275}{7} = 39.3 \text{ Gannets}$$

The calculated value of 39.3 scarcely represents all the observations in the sample. It is larger than six of the seven observations and is more than five times smaller than the other. Because the mean takes into account the value of every observation in a sample, it can be greatly distorted by a single exceptional value. Perhaps a storm out at sea drove more Gannets inshore on the 5th day. When a few exceptional values distort the mean in this way, a *resistant measure of the average*, namely, the **median**, may be more appropriate.

5.3 The median – a resistant statistic
The median is the middle observation in a set of observations which have been ranked in magnitude.

Example 5.4
Look again at the daily counts of Gannets given in Example 5.3. The observations are placed in rank order for convenience:

4	8	9	12	15	25	202
			median			

The observation 12 has three observations to the left which are smaller and three to the right which are larger. It is the sample median. Notice how it is *resistant* to the extreme observation; the median is 12 if the seventh (largest) observation is 20, 202 or 2002. The median is more *representative* of this set as a whole than the mean.

If there is an even number of observations in a sample, there is no middle value. By convention, the median is taken to be the mean of the values of the middle pair.

Example 5.5
Find the median of the following observations:

| 9.2 | 11.5 | 13.2 | 19.7 | 29.4 | 50.1 |

The median lies between the 3rd and 4th observation. It is estimated by:

$$\text{Median} = \frac{13.2 + 19.7}{2} = 16.45$$

5.4 The median of a frequency distribution

When there are several observations of the same value near to the median (as invariably there are in a frequency distribution) its calculation is a little more complicated. In a frequency distribution the median is taken to be the "$n/2$ th" observation from either end of the distribution.

Example 5.6
Calculate the median of the following nine measurements (mm):

| 2 | 3 | 3 | 4 | 5 | 5 | 5 | 6 | 10 |

1. Draw a dot diagram of the data.

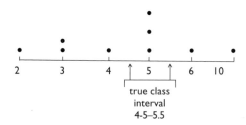

2. The median is the $n/2 = 9/2 = 4.5$, that is, the 4.5th observation from either end of the distribution.

3. Starting from the left, the cumulative frequency in the first three classes is $1 + 2 + 1 = 4$. The median is therefore the 0.5th observation in the next class. There are three observations in the next class; therefore the median is $0.5/3 = 0.167$th of the way into the class.

4. We recognise that the three observations are not *precise*: they are accurate to ± 0.5 mm. The observations fall within a class interval with lower and upper limits of 4.5 and 5.5 mm (see Section 3.9).

5. The breadth of each class interval is $5.5 - 4.5 = 1$ mm. The median is 0.167th of the way into the interval, that is, 0.167 mm. Add this to the value of the lower interval limit: $4.5 + 0.167 = 4.667$ mm. This, then, is the median.

6. Confirm the result by working down the distribution from the right. The median observation is the $n/2 = 4.5$th observation. The cumulative frequency down to the median class interval is $1 + 1 = 2$. The median is therefore the 2.5th into the interval. There are three observations in in the interval; the median, then, is $2.5/3 = 0.883$th of the way into the interval. Because the interval breadth is 1 mm, 1×0.833 is subtracted from the value of the upper interval limit, $5.5 - 0.833 = 4.667$ mm. The median checks at 4.667 mm. In practice, this would be reported as 4.7 mm.

5.5 The mode

The mode is another measure of the average. In its most common usage this measure is called the **crude mode** or **modal class**. It is the class in a frequency distribution which contains more observations than any other. The modal wing length of the Robins in Section 3.10 is 74 mm. The mode of the sample of measurements in Example 5.6 is 5 mm. The mode is the only measure of the average that can be applied to observations on *ordinal scales*.

A frequency distribution with more than one peak, or mode (not necessarily of equal height), is called a **multi-modal distribution**. Where there are *two* peaks it is a **bimodal distribution**. Fig. 5.1 shows the distribution of observations of the wing length of 240 Willow Warblers. There is one mode at the interval represented by 63 mm and another at 68 mm. In this instance the two modes correspond to the modal wing lengths of females and males. As we explain in Chapter 8, some statistical techniques depend on an assumption that data are approximately distributed in a special symmetrical way in which the mode is on the axis of symmetry. Because multi-modal distributions are not like this, it may be necessary to conduct separate analyses on discrete population categories (males, females, juveniles, etc.) for which the data are more or less symmetrical. A dot diagram usually shows if the data are multi-modal.

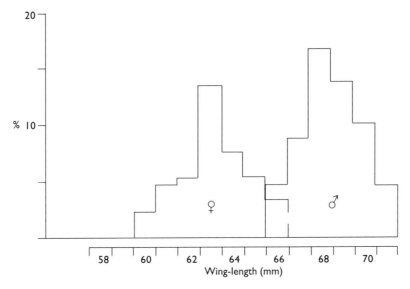

Fig 5.1 Wing length distribution of 240 adult Willow Warblers of known sex
(Ringing and Migration *4: 270*)

5.6 Relationship between the mean, median and mode

In a perfectly symmetrical distribution the mean, median and mode have the same value. In a skewed distribution (that is, a distribution which is distorted assymetrically to the left or right), the mean shifts towards the direction of the skew. In biological distributions a skew is nearly always to the right (positive skew) and so the mean is larger than the median and the mode, as Fig. 5.2 illustrates.

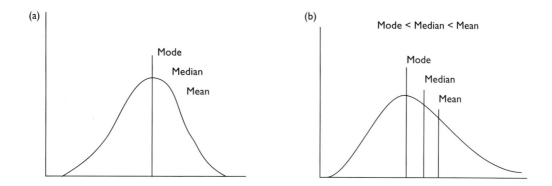

Fig. 5.2 (A) Symmetrical distribution; (B) skewed distribution.

The mean is the only one of the three measures of average which uses all the information in a set of data. That is to say, it reflects the value of each observation in the distribution. Its precisely defined mathematical value allows other statistical techniques to be based upon it. Moreover, it has the advantage of being capable of combination with the means of other samples to obtain an overall mean. This may be useful if individuals (sampling units) are rare or hard to come by.

As we have seen in Example 5.4 there are occasions when taking account of the value of every observation in a distribution gives a distorted picture of the data. The effect of introducing one or two atypically high or low observations is to pull the mean in the direction of those observations. In such cases the median provides a more realistic description of the centre of distribution than the mean. The median is especially useful, moreover, in the *preliminary* analysis of data in highlighting overall trends.

The mode is useful as a quick and approximate measure of central tendency and as an indication of the centre of distribution of observations measured on *ordinal scales*.

CHAPTER 6
MEASURING VARIABILITY

6.1 Variability

If there were no variability within populations there would be no need for statistics: a single item or sampling unit would tell all that is needed to be known about the population as a whole. It follows therefore that in presenting information about a sample it is not enough simply to give a measure of the *average*; what is also needed is information about **variability** within that sample. A dot-diagram is a qualitative method of assessing the variability in a sample, as Fig. 6.1 shows. For a quantitative analysis of variability within and between samples, we need a mathematically defined measure. The three that we describe in this chapter are the **range**, **standard deviation** and **variance**.

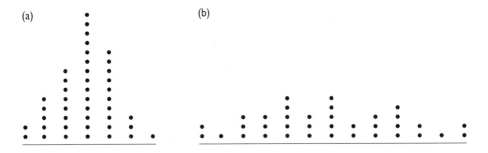

Fig 6.1 (a) Low variability (b) high variability

6.2 The range

The simplest measure of variability in a sample is the range. This indicates the highest and lowest observations in a distribution, and shows how wide the distribution is over which the measurements are spread. For continuous variables, the range is the arithmetic difference between the highest and lowest observations in a sample. In the case of counts or measurements recorded to the nearest whole unit, remember to add 1 to the difference because the range is inclusive of the extreme variations.

The range takes account of only the two most extreme observations. It is therefore limited in its usefulness because it gives no information about how observations are distributed. Are they evenly spread, or are they clustered at one end? The range should be used when observations are too few or two scattered to warrant the calculation of a more precise measure of variability, or when all that is required is a knowledge of the overall spread of the observations as, for example, when parasitologists express the range of infestation of fleas, lice, etc. on a sample of hosts.

6.3 The standard deviation

The standard deviation is the most widely applied measure of variability. It is calculated directly from all the observations of a particular variable. When observations have been obtained from every item or sampling unit in a population, the symbol for the standard deviation is σ (sigma). This is a parameter of the population and it is calculated from:

$$\sigma = \sqrt{\frac{\Sigma (x - \mu)^2}{N}}$$

where x = the value of an observation, μ = population mean, Σ = the 'sum of' and N = the number of items or sampling units in the population. It is rarely possible to obtain observations from every item or sampling unit in a population. We have therefore to be content with estimating σ from the observations in a sample. The estimate of σ is symbolised s. The value of s is not absolute, but varies from sample to sample. However, a set of values of s obtained from several samples clusters around σ. The formula for obtaining s is:

$$s = \sqrt{\frac{\Sigma (x - \bar{x})^2}{n - 1}}$$

where $\bar{x}$ is the sample mean and n is the number of observations in the sample. The term $(x - \bar{x})$ is known as the **deviation from the mean**. In cases where $\bar{x}$ is larger than x the deviation is negative; squaring the number always makes it positive. In the formula for s, note that the denominator is $(n - 1)$. The reduction by one has the effect of increasing s. Elliott (1977) describes this increase neatly as a 'tax' to be paid for using the sample mean $\bar{x}$ (a statistic) instead of the population mean μ (a parameter) in the estimation of σ. The expression $(n - 1)$ is known as the **degrees of freedom**. We explain the meaning of this expression more fully at the end of the chapter, but in estimating a standard deviation, the degrees of freedom are always one less than the number of observations in a sample. Although s is an estimate of the *population standard deviation*, it is widely referred to as the **sample standard deviation** because it is calculated from sample data.

It is possible to calculate an alternative sample standard deviation by placing n (rather than the degrees of freedom, $n - 1$) in the denominator. This, however, is a *descriptive statistic* only and has no meaning in inferential statistics. We suggest that you never use it unless you have a clear reason for doing so.

In practice, you will undoubtedly choose to derive a standard deviation directly from a scientific calculator. This usually has separate keys marked n and $(n - 1)$ for population and sample standard deviations, respectively. In the following section we show you how to calculate the standard deviation from first principles. It is important that you study this because it introduces a term called the **sum of squares** which has application in more advanced statistical methods which we touch upon in later chapters.

6.4 Calculating the standard deviation

The procedure for calculating the standard deviation from first principles is shown in Example 6.1.

Example 6.1

Calculate the standard deviation of the following 10 observations (mm).

| 81 | 79 | 82 | 83 | 80 | 78 | 80 | 87 | 82 | 82 |

1. Calculate the sample mean, $\bar{x}$:

$$\bar{x} = \frac{\Sigma x}{n} = \frac{814}{10} = 81.40 \text{ mm}$$

2. Obtain the deviations from the mean by subtracting the mean from each observation in turn; square each deviation (the squaring eliminates any minus signs):

$(81 - 81.4)^2 = 0.16$ $(78 - 81.4)^2 = 11.56$

$(79 - 81.4)^2 = 5.76$ $(80 - 81.4)^2 = 1.96$

$(82 - 81.4)^2 = 0.36$ $(87 - 81.4)^2 = 31.36$

$(83 - 81.4)^2 = 2.56$ $(82 - 81.4)^2 = 0.36$

$(80 - 81.4)^2 = 1.96$ $(82 - 81.4)^2 = 0.36$

3. Add up the 10 squared deviations: the sum of the squared deviations is 56.4. This sum, $\Sigma (x - \bar{x})^2$ is called the **sum of the squares of the deviations** or, more simply, the **sum of squares**.

4. Divide the sum of squares by one less than the number of observations:

$$\frac{\text{sum of squares}}{(n-1)} = \frac{\Sigma (x - \bar{x})^2}{(n-1)} = \frac{56.4}{9} = 6.27$$

5. The standard deviation is the square root of this value:

$$s = \sqrt{6.27} = 2.50 \text{ mm}$$

We estimate that the sample of 10 observations is drawn from a population whose standard deviation is 2.50 mm.

6.5 Variance

An important measure of variability closely related to the standard deviation is variance. Variance is used in certain parametric techniques described later. Variance is the square of the standard deviation; conversely, a standard deviation is the square root of the variance. Variance is symbolised σ^2 for a population variance, and s^2 for a variance estimated from a sample. Thus:

$$\sigma = \sqrt{\sigma^2} \qquad \text{and} \qquad s = \sqrt{s^2} \qquad \text{and} \qquad s^2 = \frac{\Sigma (x - \bar{x})^2}{(n-1)}$$

It follows that the variance is the value obtained before taking the square root in the final step of the calculation of the standard deviation. The variance of the sample in Example 6.1 is 6.27.

6.6 An alternative formula for calculating the variance and standard deviation

The method described in Section 6.4 illustrates the principle of calculating the variance and standard deviation. An algebraic rearrangement of the formula is, in practice, easier to handle. Moreover, it introduces two statistical expressions which will appear again later. The alternative formula for obtaining the sum of squares is:

$$\text{Sum of squares} = \Sigma x^2 - \frac{(\Sigma x)^2}{n}$$

The two new expressions are Σx^2 and $(\Sigma x)^2$

(i) Σx^2 is called *the sum of squares of x*. Using again the 10 measurements of Example 6.1 it is derived as follows:

| x: | 81 | 79 | 82 | 83 | 80 | 78 | 80 | 87 | 82 | 82 |

$$\Sigma x^2: \quad 81^2 + 79^2 + 82^2 + 83^2 + 80^2 + 78^2 + 80^2 + 87^2 + 82^2 + 82^2 = 66316$$

(ii) $(\Sigma x)^2$ is called the square of the sum of x and is calculated as follows:

$$(\Sigma x)^2 = (81 + 79 + 82 + 83 + 80 + 78 + 80 + 87 + 82 + 82)^2 = (814)^2 = 662596$$

Substituting in the formula for the sum of squares:

$$\text{Sum of squares} = 66316 - \frac{662596}{10} = 56.4$$

(i) The variance is obtained by dividing the sum of squares by $(n-1)$:

$$s^2 = \frac{56.4}{(10-1)} = 6.27$$

(ii) The standard deviation is the square root of the variance:

$$s = \sqrt{6.27} = 2.50 \text{ mm}$$

6.7 Obtaining the standard deviation, variance and the sum of squares from a calculator

Scientific calculators have facilities for entering observations and, by pressing appropriate keys, obtaining $\bar{x}$, standard deviation (remember to use the key marked $n-1$), Σx, and Σx^2.

If you are unsure about obtaining these numbers from your calculator, we recommend that you stop at this point and learn to do so from the instruction booklet accompanying the instrument before proceeding further.

Some calculators do not have facilities for obtaining the variance, sum of squares and $(\Sigma x)^2$ directly.

(i) To obtain the variance, first obtain the standard deviation, and then square it.

(ii) To obtain $(\Sigma x)^2$, obtain Σx and then square it.

(iii) To obtain the sum of squares, obtain the variance as in (i) and then multiply by $(n-1)$.

These three operations can be undertaken while the calculator is operating in the *standard deviation mode*.

Some makes of calculator do not permit the direct retrieval of a standard deviation for $(n-1)$. This is unfortunate because almost without exception this is what you need. To convert s_n to $s_{(n-1)}$, square the standard deviation, multiply by n, divide by $(n-1)$ and take the square root. Check the result of your attempt against the solution to Example 6.1.

6.8 The coefficient of variation CV

The standard deviation s is a measure of the degree of variability in a sample which estimates the corresponding parameter of a population. However, it is of limited value for *comparing* the variability of samples whose means are appreciably different. The standard deviation of the weight of a population of Ostriches is hundreds of grams, whilst that of a population of Goldcrests is less than a gram. When comparing variability in samples from populations with appreciably different means, the **coefficient of variation (CV)** is used. This is the ratio of the standard deviation to the mean, usually expressed as a percentage by multiplying by 100.

Example 6.2

Aspects of growth in bird chicks are investigated. The mean and standard deviation of samples of (a) eggs; (b) 4-day old chicks; (c) 10-day old chicks are recorded. Does the relative variability change with age?

(a) Eggs	$\bar{x} = 3$ g	$s = 0.54$ g	CV = 0.54/3 x 100 = 18%
(b) 4-day-old chicks	$\bar{x} = 4.5$ g	$s = 1.3$ g	CV = 1.3/4.5 x 100 = 28.9%
(c) 10-day-old chicks	$\bar{x} = 10.4$ g	$s = 4.1$ g	CV = 4.1/10.4 x 100 = 39.4%

It is clear from an inspection of the three values of CV that relative variability increases with age.

6.9 Degrees of freedom

In obtaining a sample standard deviation to estimate a population standard deviation in Section 6.3, reference is made to the number of degrees of freedom. Because the concept of *degrees of freedom* is involved in many statistical techniques, it now needs a fuller explanation.

Suppose that we are told that a sample of $n = 5$ observations has a mean $\bar{x}$ of 50 and we are then asked to 'invent' the values of the observations. We know that $\Sigma x = (\bar{x} \times n) = 250$. If the sum of the observations is 250, we have freedom of choice only for the first four numbers because the fifth must be a number (perhaps a negative number) that brings the sum to 250. By way of example, if the first four numbers are arbitrarily chosen as, say, 40, 25, 18, 130, then in order to make $\Sigma x = 250$, we have no further freedom of choice; the fifth number must be 37. Degrees of freedom (df) in our present example is one less than the number of observations. That is, df = $(n - 1)$.

Sometimes the formula for estimating a population parameter contains a value which is itself an estimate. Thus, to estimate a standard deviation, a knowledge of the mean is required. Because the value of the mean is itself estimated from a sample, this 'costs' a degree of freedom (this cost is Elliott's 'tax' referred to in Section 6.3).

The degrees of freedom are not always $(n - 1)$; they depend on the particular estimation in hand. We explain the rule for deciding the degrees of freedom for each technique as it arises. Degrees of freedom are symbolised by v (nu).

CHAPTER 7
PROBABILITY AND THE NORMAL CURVE

7.1 The meaning of probability

The mathematical theory of probability arose from the study of games of chance. That is why so many text-book illustrations of probability involve dice-throwing and coin-tossing (though we will try to stimulate your interest with slightly less mundane examples!). Probability may be regarded as quantifying the chance that a stated outcome of an event will take place. By convention, probability values fall on a scale between 0 and 1, but they are sometimes expressed as percentages. The probability that the birds you see will eventually die is $P = 1$ (certainty); the probability that you will add Dodo *Raphus cucullatus* to your "life-list" is 0 (impossibility).

Example 7.1

A ringing trainer brings back 10 bird bags to his trainee at the ringing station and says that in them there are 9 Blue Tits and a Firecrest. The trainer allows the trainee one guess to select the bag with the Firecrest in order to ring it. What is the probability that the trainee will ring the Firecrest?

$$P = \frac{\text{Number of bags with a Firecrest}}{\text{Total number of bags}} = \frac{1}{10} = 0.1 \ (10\%)$$

What is the probability that the trainee will ring a Blue Tit?

$$P = \frac{\text{Number of bags with a Blue Tit}}{\text{Total number of bags}} = \frac{9}{10} = 0.9 \ (90\%)$$

Notice firstly that the probability is equal to the *proportional frequency* of an item in the sample (Firecrest or Blue Tit, respectively) and, secondly, the sum of the separate probabilities $(0.1 + 0.9)$ equals 1. In other words, the trainee will certainly ring *something!*

7.2 Compound probabilities

Probability values may be added and multiplied. Because probability values are fractions, adding probabilities *increases* the likelihood of a stated outcome whilst multiplying probabilities *decreases* it. The decision whether to add or to multiply may therefore often be made intuitively.

Example 7.2

We extend the scenario of Example 7.1 to a rather less likely, but nevertheless feasible, situation. Imagine a large and busy ringing group that has accumulated 100 Great Tits in separate bags waiting to be processed. The ringers noted the age and sex of each bird when bagged and the age/sex composition of the "population" is as follows:

	Adult		Juvenile	
Male		Female	Male	Female
20		10	30	40

Totals:				
	30		70	100

The probability of a nominated outcome (e.g. *male, juvenile, adult female*, etc. can be worked out by a simple tree diagram in which the branches show the proportional frequencies (probabilities) of a particular event. Thus, the probability of *adult* is 30/100 = 0.3; the probability of *juvenile* is 70/100 = 0.7. Of the *adults*, 20/30 (2/3) are *male* whilst 10/30 (1/3) are *female*. Similarly, the respective probabilities for juveniles are 3/7 and 4/7.

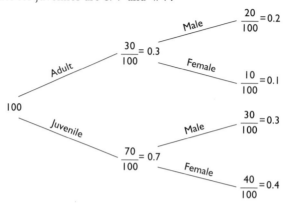

What is the probability that the first bag selected at random will contain *either* an adult or a juvenile? Clearly the probability of this outcome is greater than in which a single species is nominated:

Probability of outcome *adult*:   =   0.3
Probability of outcome *juvenile*:   =   0.7
Probability of outcome *adult* or *juvenile*   =   0.3 + 0.7   =   1

Example 7.3
What is the probability that the first bag selected at random will contain a juvenile female? In this case, since two conditions (age and sex) are stipulated, the probability is lower than for any single condition:

Probability of outcome *juvenile*   =   0.7
Probability of outcome *female*   =   (0.1 + 0.4) =   0.5
Probability *juvenile* and *female*   =   0.7 x (4/7) =   0.4

Example 7.4
If the outcome of an event is described in terms of the age and sex of the contents of a single bag selected at random, show that the sum of the probabilities for all possible outcomes is equal to 1.

Probability of outcome			
Adult male	*Juvenile male*	*Adult female*	*Juvenile female*
0.3 x (2/3) = 0.2	0.7 x (3/7) = 0.3	0.3 x (1/3) = 0.1	0.7 x (4/7) = 0.4

Total probability = 0.2 + 0.3 + 0.1 + 0.4 = 1.

A breakdown which shows the individual probabilities of all possible outcomes and which adds up to 1 is called a **probability distribution.**

7.3 Critical probability

Sometimes we wish to know the probability below which a stated outcome is considered unlikely to be the result of chance alone. Such a value is called a **critical probability**, and is of course decided arbitrarily. We may refer back to our unfortunate trainee from Example 7.1 who is expected to guess which one of a number of bird-bags contains the desired Firecrest.

Example 7.5
How ought the trainee to react if the Firecrest was correctly singled out from:
(a) one bag in 10?
(b) one bag in 20?
(c) one bag in 100?

(a) If the Firecrest was correctly singled out from one bag in ten, the trainee would undoubtedly feel a little lucky, but there should be no cause for major celebration; a probability of 0.1 means that the result would be expected from one trial in 10.

(b) The probability of singling out the Firecrest from one bag in 20 is 0.05. If this were the case, the trainee should feel extremely fortunate. The trainer might even harbour a suspicion that the trainee had narrowed down the odds a little by inspecting the bags closely.

(c) The probability of singling out the Firecrest from one bag in 100 is only 0.01. If this were the case the trainer would have very strong grounds to suppose that the trainee was not guessing, but had some insight or intuition as to which bag contained the coveted specimen.

At what point does the probability of an outcome become so low that it must be regarded as "unlikely"? Statisticians conventionally adopt three critical probability levels.

1. An outcome which is predicted to occur fewer than one trial in 20 ($P<0.05$) is considered to be *unlikely* or *statistically significant*.

2. An outcome which is predicted to occur fewer than one trial in 100 ($P<0.01$) is considered to be *very unlikely* or *statistically highly significant*.

3. An outcome which is predicted to occur fewer than one trial in 1000 ($P<0.001$) is considered to be *extremely unlikely* or *statistically very highly significant*.

These three levels of significance are often denoted *, **, and ***, respectively.

7.4 Probability distribution

Section 7.2 concludes with an example of a simple probability distribution. Probability distributions may be generated empirically, that is, by sampling and observation. Thus, recalling Fig. 4.7 which is a frequency distribution of 100 Robin wing lengths, we can convert a frequency distribution into a probability distribution by re-scaling the vertical axis and dividing by the number of observations (sample size). This has been done in Fig. 7.1. It may be seen that, for example, the probability that a Robin selected at random will have a wing length that falls within the 74 mm class is $20/100 = 0.2$. Moreover, if the individual probabilities for each frequency class are summed, the result is 1.

Why are frequency distributions useful? First, as indicated above, they allow us to estimate the probability that an event (e.g. the random selection of a Robin) will have a stated outcome (e.g. wing length falling within the 74 mm class). Second, a probability distribution may be used to generate a distribution of **expected frequencies**. Just as a probability is estimated by dividing a frequency of a particular observation by the total number of all observations, an expected frequency may be produced by *multiplying* a probability by the number of observations:

Expected frequency = (estimated probability) x (number of observations).

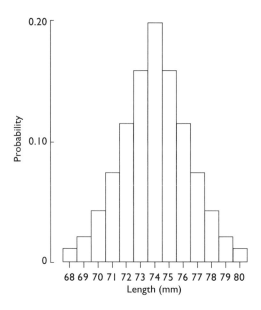

Fig. 7.1 Probability distribution of leaf length (±0.5mm) estimated from a sample of 100 leaves

Example 7.6

Using the probability distribution worked out in Example 7.4, what is the *expected* frequency of each outcome in terms of age and sex in a sample size of 240 Great Tits observed in a separate habitat?

Expected frequency of *adult male*	=	0.2 x 240	=	48
Expected frequency of *juvenile male*	=	0.3 x 240	=	72
Expected frequency of *adult female*	=	0.1 x 240	=	24
Expected frequency of *juvenile female*	=	0.4 x 240	=	96

Total expected frequency   =   240

In practice, the obvious thing to do next is to compare the expected frequencies of each outcome with actual or observed frequencies. If the observed frequencies disagree with those expected, there could be a basis for exploring factors which underlie the discrepancy.

7.5 The Normal Distribution

In Fig. 4.7 we showed the distribution of 100 Robin wing lengths measured to a precision of ± 0.5 mm. Because measurements of length are on a continuous scale, they may be made with increased precision to, say, ± 0.05 mm. Each frequency class then corresponds to increments of 0.1 mm and, if the number of observations is increased accordingly, the steps in the histogram become more gradual as shown in Fig. 7.2(a). As increasing numbers of observations are obtained with increasing degrees of precision, the histogram verges towards a smooth, symmetrical bell-shaped curve as shown in Fig. 7.2(b).

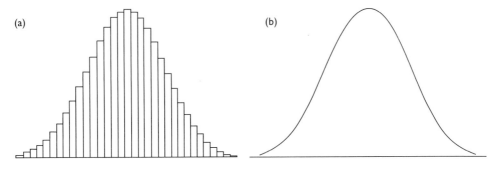

(a)

(b)

Fig. 7.2 Gradation of the histogram (a) into the normal curve (b)

The mathematician Gauss discovered that distributions of this shape estimated from large samples of measurements drawn from a single population often agree well with an intimidating mathematical formula which we do not need to print here. The important feature of the formula is that it is governed by two constants, namely μ (the population mean) and σ (the population standard deviation). These are said to be the **parameters of the distribution**, and once they have been estimated for a particular population, the shape of its distribution curve can be worked out using the formula. The name of this distribution is the **normal distribution**. Usually we do not know the values of μ and σ and have to estimate them from a sample as $\bar{x}$ and s. If the number of observations in the sample exceeds about 30, $\bar{x}$ and s are considered to be reliable estimates of the parameters.

The main features of the normal distribution curve are shown in Fig. 7.3. We see that the curve is at its highest where an observation x along the base-line is equal to the population mean. Fig. 7.3(a) shows curves with similar μ but different σ; Fig. 7.3(b) shows two curves of different μ and similar σ. In reality, these parameters may vary continuously, generating an infinite variety of normal curves. It is because a single formula governs them all that they have certain properties in common which can be exploited by statisticians.

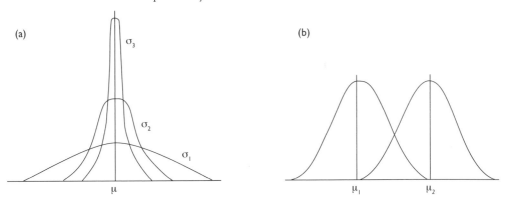

(a)

σ_3

σ_2

σ_1

μ

(b)

μ_1

μ_2

Fig. 7.3 Normal curves. (a) Similar μ and different σ; (b) different μ and similar σ.

7.6 Some mathematical properties of the normal curve

A normal curve is symmetrical, with the axis of symmetry passing through the baseline where $x = \mu$. Theoretically, the two tails of the curve never actually touch the base line; rather they continue to approach it over an infinite distance. In real life, however, the tails are never very long – we do not come across members of a single population that are both astronomically large and microscopically small.

On each side of the curve is a point of inflection where the direction of the curve changes from concave to convex. It so happens that if a vertical line is dropped down from the point of inflection it cuts the baseline at a distance on either side of the central axis equal to the standard deviation. The distance can be used as a standard unit to divide up the baseline (the *x*-axis) into equal segments, as shown in Fig. 7.4

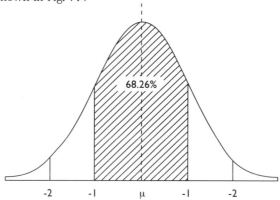

Fig. 7.4 Normal curve with standard deviations.

If the vertical axis of the distribution is re-scaled by dividing by the number of observations, it effectively becomes a probability distribution or, strictly, a **probability density**. That is, the total probability encompassed by the area under the curve is 1 (100%). One of the mathematical properties of the normal curve is that the area bounded by one standard deviation on each side of the central axis (that is, $\pm\sigma$) is approximately 68.26% of the total area. This means that about 68% of observations drawn randomly from a normally distributed population will fall within ±1 standard deviation from the mean. The other 32% fall outside these limits, 16% above one standard deviation and 16% below one standard deviation. In other words, there is a probability of $P = 0.68$ that a single observation drawn at random will fall between $\pm\sigma$.

By extending these limits to two standard deviations ($\pm2\sigma$) the proportion of observations that will be included within them is increased to 95.44%. Similarly, 99.74% of observations fall within $\pm3\sigma$). These are equivalent to probabilities of 0.9544 and 0.9974. In practice, values of 0.95 and 0.99 are more convenient to deal with. It can be calculated that these probability values fall at $\pm1.96\sigma$ and $\pm2.58\sigma$, respectively. It follows that the probability of a random observation falling outside these limits is 0.05 and 0.01. It will be recalled from section 7.3 that these are the critical probability values for assessing whether the outcome of a stated event is *unlikely* or *very unlikely*. Therefore, we may use the properties of the normal curve to assess the likelihood of certain outcomes, as we show in the next section.

7.7 Standardising the normal curve

Any value of an observation *x* on the baseline of a normal curve can be standardised as a number of standard deviation units the observation is away from the population mean, μ. This is called a *z*-score. To transform *x* into *z* apply the formula:

$$z = \frac{(x - \mu)}{\sigma}$$

If μ is larger than *x*, then *z* is negative.

It is often the case that we do not know the values of μ and σ. In samples of more than about 30 observations $\bar{x}$ and s are considered to be good estimates of μ and σ and so the formula for z is given by:

$$z = \frac{(x - \bar{x})}{s}$$

By **standardising** an observation x into a z-score, we can relate it to the properties which apply to all normal curves. Thus, if the calculated value of z is larger than 1.96, then the probability of such an observation being drawn at random is less than 0.05. According to our definition in Section 7.3, this is regarded as *unlikely* or **statistically significant**.

Example 7.7

On the basis of a very large sample, the population mean length (μ) of Coal Tit eggs was estimated to be 15.2 mm and σ estimated to be 0.55 mm. Is it likely that a single egg of length 16.4 mm found in a collection belongs to the same population?

$$z = \frac{(x - \mu)}{\sigma} = \frac{(16.4 - 15.2)}{0.55} = 2.18$$

The calculated value of z is larger than 1.96 which is equivalent to to $P<0.05$. We say that the observation is statistically significant. However the calculated z-score does *not* exceed the value of 2.58 and so it is not *statistically highly significant*. We conclude that the egg probably does not come from the same population from which the original measurements were taken.

This computation and our resulting conclusion is an example of a simple statistical test. We will recall it when we outline the principles of statistical testing in Chapter 10.

7.8 Two-tailed or one-tailed?

In Section 7.6 we said that under a normal curve 95% of observations are contained by $\mu \pm 1.96\sigma$, and that the residual 5% is divided equally outside these limits in both tails of the distribution. This is shown in Fig. 7.5. There is an alternative way of apportioning 95% of the observations and that is by excluding the residual 5% in a single tail (either the *upper* tail or the *lower* tail). In this case the cut-off point occurs at the smaller value of 1.65 standard deviations on whichever side of the distribution the 5% is excluded. This is shown in Fig. 7.6.

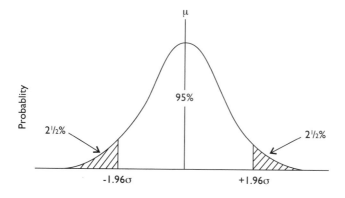

Fig. 7.5 The normal curve with 5% in two tails.

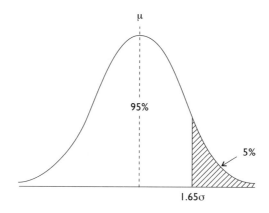

Fig. 7.6 The normal curve with 5% in upper tail.

Sometimes we are able to say that if a random observation probably does *not* belong to a population of specified mean and standard deviation, then it *must* belong to one that has a mean which is specified as being larger or, alternatively, to one that has a mean which is specified as being smaller. In this case, the critical value of z is set at the lower value of 1.65 and the test is said to be *one-tailed*. As we warn in a later chapter, there are dangers associated with one-tailed tests, and it is safer to proceed on the assumption that a test is not one-tailed unless there are clear reasons for supposing otherwise.

7.9 Small samples: the *t*-distribution

In the normal distribution the z-scores of 1.96 and 2.58 indicate the limits on either side of a population mean within which 95% and 99% of all observations will fall. These values are based on the assumption that we know the values of the population parameters μ and σ. Usually we do not know the values of μ and σ and are obliged to estimate them from a sample. We can only be confident that a sample mean and standard deviation are reliable estimates of the population parameters if the sample is large. When the sample is small, we are less confident. To compensate for the uncertainty the values 1.96 and 2.58 are increased, that is, set further out from the mean and the symbol z is replaced by the symbol t. As the sample size decreases, so too does the degree of certainty. The values of t must therefore increase accordingly. Thus, in addition to the mean and standard deviation parameters, t-distributions are also dependent on sample size. However, it is not n which determines t directly, but $(n - 1)$, that is, the degrees of freedom. There are mathematical techniques available for working out the complete probability distributions of t for any number of degrees of freedom. In practice we usually need to know only the values which correspond to $P = 0.05$ and $P = 0.01$ for a particular sample size, that is, the values which correspond to the z-scores of 1.96 and 2.58 in large samples.

The values are looked up in tables which are given in Appendix 1.

Example 7.8
What is the critical value of t for a sample of 12 observations?

Enter the table in Appendix 1 at $12 - 1 = 11$ df. Under the heading 0.05 in the "two-tailed test" we find the value 2.201. This is a little larger than the corresponding value of z, namely 1.96.

Examination of the t-table for $P = 0.05$ shows that as the degrees of freedom increase, the value of t decreases until it converges towards the value of z (1.96) when $n = \infty$ (infinity). Above 30 df, however, the relative change in the value of t becomes very small with increasing sample size; it is very slightly larger than 2.0. Thus, when the sample size exceeds about 30, the difference between z and t may often be ignored. In published accounts of their work many biologists often settle for the arithmetic convenience of $\bar{x} \pm 2$ standard deviations as the critical limit unless samples are very small.

7.10 Are our data 'normal'?

Many parametric statistical techniques that we describe depend on the properties of the normal curve. They usually assume that samples have been drawn from populations which are normally distributed. Sometimes, if samples are very small, it is hard to know if the parent population is normal, in which case **distribution free** (**non-parametric**) techniques are appropriate.

Because the normal curve is a continuous, smooth curve, samples which comprise units of *count* data can never, in theory, be normal. Because counts are integral whole numbers, a frequency distribution will always have distinct "steps" and will never conform to a smooth curve. However, error due to the lack of continuity (that is, due to the 'steps' in the distribution) need not be serious if the distribution is fairly symmetrical. Then we may regard the distribution as being *approximately normal.* Errors are more likely to be serious when distributions are badly *skewed* such as we often find in count data, for the reason that these distributions are automatically truncated at zero, thus restricting the capacity for symmetry (Fig. 8.1) Data of these kind should not be used in parametric tests without first checking that they are are approximately normal, or *transforming* the data to *normalise* them. Chapter 8 considers methods of transforming data.

Even samples of "measurement" data may depart seriously from normality under some circumstances. Many organisms are polymorphic, for example sexually dimorphic. Samples of measurements of individuals from such populations will be multi-modal with each mode probably corresponding to a population class, for example sex or age class. Before any sensible statistical analysis can be undertaken the individual classes, if possible, should be separated out and separate analyses conducted on each class individually (see Section 5.5).

There are sophisticated methods for testing the normality of data but, in practice, a much simpler method suffices: simply plot out the data and see if they *look* normal! As a back-up, calculate the mean and standard deviation of the sample and see if about 70% of the observations fall within the interval $\bar{x} = \pm s$.

Example 7.9
Three different samples with 10 observations in each are shown below. Do they appear to have been drawn from normally distributed populations?

(a)

x	f
6	1
7	1
8	3
9	2
10	2
11	1

(b)

x	f
5	3
6	3
7	2
8	1
9	1

(c)

x	f
5	3
6	2
7	1
8	2
9	2

(a) A dot-diagram shows:

```
            .
      .  .  .
.  .  .  .  .  .
_____
6   7   8   9  10  11
```

$\bar{x} = 8.6$, $s = 1.5$; $\bar{x} \pm s$ is 7.1 to 10.1 which contains 7/10 observations. Although the distribution is slightly skewed, the observations are scattered on either side of a mode, and 7 out of the 10 observations fall within $\bar{x} + s$. There is no reason to doubt that the sample is drawn from a normal population.

(b) A dot-diagram shows:

```
   .  .
   .  .  .
   .  .  .  .  .
_____
5   6   7   8   9
```

$\bar{x} = 6.4$, $s = 1.35$; $\bar{x} \pm s$ is about 5.05 to 7.55 which contains 5/10 observations. In this case there is a perceptible skew in the data. Only half the observations fall within $\bar{x} \pm s$. We doubt that the sample is drawn from a normally distributed population.

(c) A dot diagram shows:

```
   .
   .  .     .  .
   .  .  .  .  .
_____
5   6   7   8   9
```

$\bar{x} = 6.8$; $s = 1.62$; $\bar{x} \pm s$ is 5.18 to 8.42 which contains 5/10 observations. In this case there is marked bimodalism. We doubt that the sample is drawn from a normally distributed population.

CHAPTER 8
DATA TRANSFORMATION

8.1 The need for transformation

In Chapter 7 we explain how an infinite variety of normal curves may be generated by varying the values of the two parameters μ and σ. In every one of such curves a constant proportion of its area is enclosed by $\mu \pm$ a given number of standard deviation units. A number of important statistical techniques depend upon this property and their correct application assumes that samples are drawn from populations which are normally distributed. Many samples of 'count' data are strongly skewed and do not conform to normal distributions. This is the case when the objects being counted are clumped, or *aggregated* and some sample units contain few (0, 1) objects whilst the occasional one contains a large number. When a distinct skew exists there is a risk of error if the data are not first *normalised* by transformation.

There is a second reason why data transformation may be necessary. Parametric techniques which compare the means of two or more samples assume that the variances of each sample are so similar that differences between them may be ignored. Samples of observations which are counts are unlikely to meet this condition. As noted in Section 7.10, distributions of count data are automatically truncated at zero. The truncation is more marked when the mean is small or near to zero. As a sample mean gets larger, there is scope for more spread on either side of the mean, and so the variance is accordingly larger (Fig. 8.1). We say the *the variance is dependent on the mean*. It is a fortuitous mathematical fact that transformation techniques which normalise data also tend to remove the dependency of the variance upon the mean. Transformation is said to *stabilise the variance*.

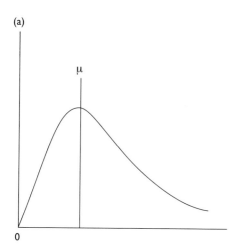 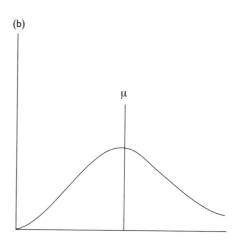

Fig. 8.1 (a) Small μ: spread of observations below μ is truncated by zero.
(b) Larger μ : spread of observations below μ is less restricted.

Transformation means simply the conversion of the raw values of each observation x into a mathematical derivative. The tests are then performed upon the transformed data rather than the raw observations. The three most widely used transformations in biology are logarithmic, square root and arcsine transformations. We now describe the application of each in turn.

8.2 The logarithmic transformation

The logarithmic transformation is appropriate when the variance of a sample of count data is considerably greater than the mean, as will be the case when the objects we are counting are *aggregated*. In this transformation, each observation x is replaced by the log of itself: that is, replace x by log x.

Example 8.1

In order to estimate the number of Puffin burrows spread all over a flat 1 ha island, an expedition marks out 88 squares 2 m x 2 m (sampling units), the positions of which are selected randomly. The number of burrows recorded by careful counting in each square are set out in the frequency table below: thus, 2 squares had 3 burrows, 4 had 4, 6 had 5, and so on.

x:	3	4	5	6	7	8	9	10	11
log x:	0.447	0.602	0.699	0.778	0.845	0.903	0.954	1.00	1.04
frequency f:	2	4	6	8	11	11	10	8	6

x:	12	13	14	15	16	17	18	19	20
log x:	1.08	1.11	1.15	1.18	1.20	1.23	1.26	1.28	1.30
frequency:	5	4	3	3	2	2	1	1	1

Figure 8.2 (a) shows a frequency bar graph of the raw data. Notice the positive skew, which is due to the fact that the burrows are *aggregated*. The distribution is clearly far from being normal. In Fig. 8.2 (b) the distribution is shown with the horizontal axis re-scaled (*transformed*) by replacing each value of x (2, 3, 4, 5, ... 20) by its logarithm. Notice how the right hand tail has been squashed up, making the overall shape more symmetrical – much more like the shape we expect to see in a normal distribution. The transformed shape, of course, can never be *exactly* normal (remember that a normal curve is continuous, and we are dealing here with discrete counts) but it has been *normalised* to an extent that the data may be used in parametric techniques without risk of serious error. Also, the dependence of the variance upon the mean has been eliminated.

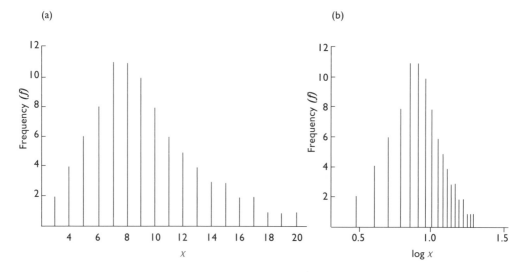

Fig. 8.2 (a) Untransformed observations; (b) logarithmically transformed observations.

We can examine the effectiveness of the transformation from the sample statistics:

Untransformed	Transformed
$n = 88$	$n = 88$
$\bar{x} = 9.35$	$\bar{x} = 0.935$
$s = 3.77$	$s = 0.18$
$s^2 = 14.2$	$s^2 = 0.03$

We see that the variance s^2 of the untransformed counts is higher than the mean. Transformation brings the value of the variance below that of the mean which is good evidence that the transformation is effective.

It is not unusual when counting things to find that some sampling units contain none of the objects being counted i.e. $x = 0$. If this should be the case, taking the logarithm of 1 is impossible (try it on your calculator if you are in doubt!). The problem is overcome by adding 1 to *all* values of x before taking the logarithm so that x is replaced by $\log(x + 1)$. Zero counts then become $\log 1 = 0$; 1 becomes $\log 2 = 0.30$, and so on.

8.3 The square root transformation

Although many objects are dispersed in nature in an aggregated fashion, we sometimes encounter items which are more or less *randomly* dispersed. We can recognise this when samples of data have a variance which is approximately *equal* to the mean. Truly random dispersions are rather rare in nature, but where they do occur, a *square root transformation* is appropriate. Lone seabirds passing a headland on passage might consititute such a dispersion; so too do lightning strikes and goals in football matches (read Moroney 1956, pp. 96–107 for a discussion on this).

Example 8.2
A seawatcher counts the number of Arctic Skuas flying south past a headland in sampling units of 15 minutes. The observations are:

x	4	6	3	3	9	4	6	2	5	2
$\sqrt{x}$	2	2.45	1.73	1.73	3	2	2.45	1.4	2.24	1.4

What is the mean of the sample?

Using a calculator the sample mean $\bar{x}$ is found to be 4.4, the standard deviation s is 2.17 and the variance s^2 is 4.71. The variance is very similar to the mean and a random dispersion is assumed. In this case a suitable transformation is to replace each observation x with $\sqrt{x}$, as shown in the table above.

The mean of the transformed counts is $\Sigma x/n = 2.04$, $s = 0.507$ and s^2 is 0.26. The variance of the transformed counts is now well below the mean and we accept that the transformation is effective.

If there are zero counts (i.e. there are some sampling units where $x = 0$) it is conventional to add 0.5 to *all* values of x before taking the square root; that is, replace x by $(x + 0.5)$.

We have identified instances where the variance is *larger* than the mean, and where it is about *equal* to the mean of samples of count data. Logic suggests there is a third possibility, when the variance is less than the mean. This occurs when the objects we count are dispersed *evenly* so that there are similar numbers of the objects in each sampling unit. This would be the case when nests are counted in a sample of quadrats marked out in a Gannet colony. The *square root* is considered to be, in practice, an adequate transformation for data like these.

8.4 The arcsine transformation

The arcsine transformation is appropriate for observations which are *proportions*. We said in Section 8.1 that distributions of count data cannot be symmetrical because the left hand tail of the distribution is truncated by the theoretical minimum of zero. In distributions which are proportions, *both* tails are truncated because all values must lie on a scale with absolute limits of 0 and 1. Errors which might arise are greatest if the observations are grouped at one end of the scale (say, close to 0.1 or 0.9) and are least if they are near the middle.

The arcsine transformation requires two steps. First, obtain the square root of x. Second, find the angle whose sine equals this value. On most calculators it is obtained by pressing the $\sin^{-1}$ (inverse sine) key.

$$\text{Thus, if} \quad x \ = \ 0.25$$
$$\sqrt{x} \ = \ 0.5$$
$$\sin^{-1} 0.5 \ = \ 30°$$

Therefore, arcsine $0.25 = 30°$.

Notice that the scale of the transformation is *angular degrees*. If observations are percentages, divide each by 100 before transforming.

Example 8.3

In a study of the distribution of bridled Guillemots, an observer determines the proportion of bridled birds in a set of sampling units in a colony on Fair Isle. A sampling unit is defined here as a discrete portion of the colony which can be clearly observed. The observations obtained are:

| 0.25 | 0.31 | 0.21 | 0.24 | 0.30 | 0.29 | 0.22 |

The mean of the proportions is $\Sigma \ x/n = 0.26$ and the standard deviation is 0.040. Because it is known that proportions are not normally distributed, the arcsine transformation is applied. Transforming x to arcsine x:

x:	0.25	0.31	0.21	0.24	0.30	0.29	0.22
Arcsine x:	30.0	33.8	27.3	29.3	33.2	32.6	28.0

Using a scientific calculator we find that the mean of the transformed observations is $30.6°$ and $s = 2.60°$.

8.5 Back-transforming transformed numbers

Usually we transform data to normalise them and to stabilise the variance so that they may be used in parametric techniques which require these conditions. Nearly always a test is performed upon the transformed data and inferences drawn from the outcome of the test. In certain techniques, however, the object is to estimate the number of objects in some population. In such cases, a transformed number may be difficult to interpret – many ornithologists would be hard pressed to think of the Guillemots in Example 8.3 in terms of angular degree units! An example is given in Section 9.5 where a number is restored to the original scale by **back-transformation**. This is simply the reversal of transformation.

(a) *Log x transformation*
   To back-transform 1.380, obtain the antilog.
   Antilog $1.380 = 24$.

(b) *Log (x + 1) transformation*
    To back-transform 1.0414, take the antilog and subtract 1.
    (Antilog 1.014) − 1 = 10.

(c) *Square root transformation*
    To back-transform 3.464, obtain the square.
    $3.464^2 = 12$.

(d) *Square root (x + 0.5) transformation*
    To back-transform 3.937, obtain the square and subtract 0.5.
    $3.937^2 - 0.5 = 15$.

(e) *Arcsine transformation*
    To back-transform 28.2°, obtain the sine and square.
    $(\sin 28.2)^2 = 0.2233$.

8.6 Is data transformation really necessary?

Users of statistics are sometimes accused of 'fiddling' or 'massaging' data if their observations do not at first support their predictions. Is the transformation of data by the methods we have outlined a dubious form of statistical massage? The answer is a firm *no!* First, data transformation is a well-established mathematical practice; the units of acidity (pH), noise (decibels) and seismic activity (Richter units) are three well-known examples of transformed scales. Second, in statistics the transformation of data is undertaken *a priori*; that is to say, it is planned in advance as part of the statistical technique. Transformation should not be used as an afterthought (*a posteriori*) to change an initial 'unfavourable' result. If it is, then this *is* cheating! Transformation merely ensures that a particular statistical method can be validly applied.

Although the process of transforming data is not difficult, with large data sets it can be laborious and repetitive even with a calculator. The question therefore arises as to whether it is worth the effort. If no more is required than a simple statement of the mean and standard deviation or variance, transformation would not be expected. But we reiterate, for valid use some statistical techniques require that the data are normal and with stable variance. If data are badly skewed, and therefore do not meet these conditions, then errors of unpredictable magnitude may arise if the data are not first transformed.

However, there is an alternative to data transformation. The non-parametric 'distribution-free' techniques we describe later in the Guide do not assume or require normally distributed data and can be used safely without transformation.

CHAPTER 9
HOW GOOD ARE OUR ESTIMATES?

9.1 Sampling error

Usually we obtain a sample in order to derive a statistic (a mean, for example) from the observations in the sample. The sample statistic gives an estimate of the corresponding parameter; thus, $\bar{x}$ estimates μ.

Intuitively, we expect that large samples provide more reliable estimates than small samples and, conversely, that small samples are less reliable than large ones. Several small samples drawn from the *same* population generally provide different values of the same statistic, yet they are all estimates of the same population parameter. The variation between these individual estimates is called **sampling error**.

Sampling error arises because some samples have, by chance, more than a 'fair share' of larger units whilst others have more than a 'fair share' of smaller units. Sampling error is *not* a mistake or error due to an observer; rather it reflects the random scatter inherent in any sample. In a collection of small samples drawn from the same population, some values of a statistic underestimate the parameter and others overestimate it. The way in which sample statistics cluster around a population parameter is called the distribution of the statistic or, sometimes, the **sampling distribution**. Distributions of sample statistics conform to mathematical principles which allow us to state the confidence we may place in estimates of particular population parameters.

9.2 The distribution of a sample mean

The 100 observations of Robin wing lengths in Section 3.10 are presented again in Table 9.1 together with the mean of each row. The grand mean of the whole sample has already been worked out as 74.00 mm. Scrutiny of the 10 row means reveals that they vary, ranging from 72.7 mm to 74.7 mm, with not a single one equal in value to the grand mean.

Table 9.1 100 Robin wing length measurements

Wing length (mm)										Mean $\bar{x}$
76	73	75	73	74	74	74	74	74	77	74.4
74	72	75	76	73	71	73	80	75	75	74.4
68	72	78	74	75	74	69	77	77	72	73.6
72	76	76	77	70	77	72	74	77	76	74.7
78	72	70	74	76	72	73	71	74	74	73.4
75	79	75	74	75	74	71	73	75	73	74.4
75	70	73	75	70	72	72	71	76	73	72.7
74	76	74	75	74	76	75	75	73	73	74.5
78	74	73	75	74	73	72	76	73	76	74.4
74	71	72	71	79	78	69	77	73	71	73.5

We can group the row means into a frequency distribution just as we did for the observations themselves into a frequency distribution in Table 3.2.

Frequency class	Tallies	Frequency f
72.5 – 73.4	11	2
73.5 – 74.4	111111	6
74.5 – 75.4	11	2

Although there are only 10 of these sub-set means, a clear, symmetrical pattern begins to emerge. Had we had more sub-set means drawn from a larger sample, their distribution would resemble that shown in Fig. 7.2 (a). Because the sample mean is a continuous variable, the distribution verges towards a smooth continuous curve, that is the curve of normal distribution. This conclusion follows from a fundamental principle in statistics:

The Central Limit Theorem states that the means of a large number of samples drawn randomly from the same population are normally distributed and the 'mean of the means' is the mean of the population.

This normal distribution has its own standard deviation, that is, a **standard deviation of sample means**. The standard deviation of a set of sample means is given its own name: the **standard error of the mean**.

It is important to note that the Central Limit Theorem makes no assumptions about the underlying distribution of the population from which the samples are drawn. That distribution does *not* have to be normal; it may be symmetrical, skewed or bimodal; observations may be continuous or discrete. Whichever is the case, the means of a large number of samples are approximately normally distributed around the population mean, μ.

The properties of the normal curve described in Chapter 7 hold true for a normal distribution of sample means. Thus about 68% of a large number of sample means fall within ± 1 standard error (S.E.) of the population mean μ. The converse of this is also true: we are similarly confident (68%) that a population mean μ falls within ± 1 S.E. estimated from a *sample* mean $\bar{x}$.

In practice we do not estimate a standard error by looking at the spread of sub-sets of sample means. Instead, it may be calculated from the observations of a sample by:

$$\text{S.E.} = \frac{\text{sample standard deviation}}{\sqrt{\text{number of observations}}} = \frac{s}{\sqrt{n}}$$

Example 9.1
Calculate the standard error of the mean of the 100 Robin wing length measurements for which we have already worked out the mean to be 74.00 mm and standard deviation to be 2.34 mm.

$$\text{S.E.} = \frac{s}{\sqrt{n}} = \frac{2.34}{\sqrt{100}} = \frac{2.34}{10} = 0.234$$

In plain language this means: the mean wing length of the sample is 74.00 mm and the standard error of the mean is ± 0.234 mm. We are therefore 68% confident that that the mean of the *population* lies between 74.00 + 0.234 (= 74.234 mm) and 74.0 – 0.234 (= 73.766 mm). This is, of course, more informative than the standard deviation (s) because it indicates how close to the sample mean is likely to be to the population mean which is what we seek to estimate.

9.3 The confidence interval of the mean of a large sample
The standard error of the mean gives us an indication of how good an estimate a sample mean, $\bar{x}$, is of a population mean, μ. Thus, we are 68% confident that μ lies within ± 1 S.E. of $\bar{x}$. However,

68% is a rather low level of confidence; we usually want to be surer that a population mean lies between indicated limits. To meet this need, 95% or 99% limits are generally used. These can be obtained by multiplying the standard error by the appropriate z-score as follows:

- we are 95% confident that a population mean falls within ± 1.96 S.E. of a sample mean

- we are 99% confident that a population mean falls within ± 2.58 S.E. of a sample mean.

1.96 and 2.58 are the same z-scores that are used in describing the properties of the normal curve and are valid only in samples containing 30 or more observations. The intervals $\bar{x} \pm 1.96$ S.E. and $\bar{x} \pm 2.58$ S.E. are called the **95%** and **99% confidence intervals**, respectively. The adjustments made in the case of smaller samples are described in the next section.

Notice the use of the word *confidence*, rather than *probability*. In other words, the interval is likely to catch the population mean in 95 trials in 100; we can not make statements about the probability of a *single* outcome.

Example 9.2
Estimate the 95% confidence interval of the mean of the 100 Robin wing length measurements given in Section 9.2.

We have determined in Example 9.1 that the mean of the sample is 74.00 mm and the standard error is ± 0.234 mm. The 95% confidence interval is therefore $74.00 \pm (1.96 \times 0.234) = 74.00 \pm 0.459$ mm. This means we are 95% confident that the population mean lies between 74.459 mm and 73.541 mm.

Notice that because n, the number of observations, is in the denominator of the equation for estimating the standard error (and hence the confidence interval) the value of the standard error (and breadth of the confidence interval) gets smaller as n gets larger. This is a mathematical expression of the statistical axiom that the larger the sample size, the greater is the reliability of an estimate of a population parameter.

9.4 The confidence interval of the mean of a small sample
In calculating the standard error of the mean (and hence a confidence interval) the standard deviation is used. Strictly, this should be the population standard deviation, σ. In large samples we are confident that the sample standard deviation s is a reliable estimate of σ; we are less certain however in the case of small samples. We therefore need to apply a correction factor to compensate for the uncertainty in small samples. That factor should become larger as the sample becomes smaller. A suitable correction factor is t, as described in Section 7.9.

Example 9.3
Calculate the mean, with 95% confidence limits, of a sample of observations of Dunnock weights (g):

| 19.4 | 21.4 | 22.3 | 22.1 | 20.1 | 23.8 | 24.6 | 19.9 | 21.5 | 19.1 |

Using a scientific calculator it is found that $\bar{x} = 21.42$ and $s = 1.84$. The standard error is:

$$\text{S.E.} = \frac{s}{\sqrt{n}} = \frac{1.84}{\sqrt{10}} = 0.582$$

In a small sample we do not use the z-score of 1.96 to obtain the confidence interval but the value of t is found in tables against the appropriate number of degrees of freedom $(n-1)$. Thus:

$$95\% \text{ confidence interval} = \bar{x} + (t \times \text{S.E.})$$

In Appendix 2 we find that for $(10-1)$ degrees of freedom at $P = 0.05$ (95%), $t = 2.262$. The 95% confidence interval is therefore:

$$21.42 \pm (2.262 \times 0.582), \text{ i.e. } 21.42 \pm 1.32 \text{ g}$$

Thus, we are 95% confident that the mean weight of the population of Dunnocks from which the sample was drawn lies between 22.74 g (upper limit) and 20.1 g (lower limit). If we had used the z-score of 1.96 instead of the t-score 2.262, the limits would have been 22.56 g and 20.28 g – an important reduction in range of about a third of a gram.

9.5 The confidence interval of the mean of a sample of count data

Because the Central Limit Theorem states that a set of sample means drawn from any single population is normally distributed about the population mean, μ, the calculation of a 95% confidence interval of the mean of a large sample of observations of counts is exactly the same as described in Section 9.3. However, because samples of count data are often drawn from samples which are greatly skewed, the application of the factor t may not be a sufficient correction in the case of a small sample. As a precaution, the confidence interval about the mean of a small sample of count data should be calculated upon transformed observations. We have described various transformation techniques in Chapter 8, and because the logarithmic is the most commonly applied we choose this one as our example to illustrate the technique for estimating a confidence interval.

Example 9.4

Returning to our flat island covered with Puffin burrows, an ornithologist marks out twelve 10 m² quadrats (squares about 3.2 x 3.2 m) at randomly selected positions and counts the number of Puffin burrows in each quadrat. The number of burrows counted are:

| 0 | 0 | 4 | 2 | 58 | 1 | 0 | 22 | 5 | 7 | 17 | 1 |

The mean of the counts $\bar{x}$ is 9.75 and the variance s^2 is 281. Because the variance is very much larger than the mean, a logarithmic transformation is appropriate (evidently the burrows are dispersed in clumps). Note that there are some zero observations; we therefore apply the transformation $\log(x + 1)$. The transformed data become:

| 0 | 0 | 0.699 | 0.477 | 1.771 | 0.301 | 0 | 1.362 | 0.778 | 0.903 | 1.255 | 0.301 |

The mean of the transformed counts is 0.654 and the standard deviation is 0.584. The standard error of the mean is:

$$\text{S.E.} = \frac{s}{\sqrt{n}} = \frac{0.584}{\sqrt{12}} = 0.169$$

The 95% confidence interval of the transformed counts is therefore:

$0.654 \pm (t \times 0.169)$ where $t\,(0.05)$ for 11 df is 2.201.

The interval is:

$$0.654 \pm (2.201 \times 0.169) = 0.654 \pm 0.372$$

Back-transforming by taking antilogs and subtracting 1, this becomes

$$3.508 \overset{x}{\underset{\cdot}{\div}} 1.355$$

[**Note:** a number which is added and subtracted as a logarithm becomes multiplied and divided when antilogged].

The number 3.508 is the *derived mean* (usually symbolised $\bar{y}$) and is considerably smaller than the arithmetic mean of the raw data, 9.75. It is known as the **geometric mean** and, if used in population estimates, results in corresponding underestimation. Although there are mathematical correction factors available, a satisfactory practical approximation is to use the **arithmetic mean** as the estimate of μ but the *limits derived by transformation* for setting the confidence interval. Thus, in the present example, we are 95% confident that if we were able to count the burrows in every possibly quadrat that could be marked out on the island (i.e. the statistical population) the mean number of burrows per quadrat would fall between:

9.75 x 1.355 = 13.21 (upper limit)   and   9.75 ÷ 1.355 = 7.196 (lower limit).

How many burrows on the island?

If we say that the area of the island is 5000 m², the crude estimate of the total number of burrows is the sample mean (untransformed) x N (the total number of all possible sampling units). That is, 9.75 x 500 = 4875 burrows. We apply the confidence interval calculated from the *transformed data*, that is, 13.21 x 500 = 6605 (upper limit) and 7.196 x 500 = 3598 (lower limit). Notice that the confidence interval is not spaced symmetrically above and below the mean. This is, of course, consistent with a very skewed distribution.

9.6 The difference between the means of two large samples

Ornithologists often compare the means of some variable in two samples. To know how much heavier, larger, faster, or parasite-infested one sample is than another allows us to make inferences about differences between the populations from which the samples are drawn.

Example 9.5
An ornithologist studying Starlings occupying a communal winter roost observed that samples caught at the edge of the roost are on average lighter in weight than those caught near the centre. It is speculated that lighter, less "dominant" individuals are relegated to the less favourable periphery. The starlings occupying the "edge" and the "centre" are regarded as belonging to different *statistical* populations. In order to quantify the difference in weight, a large sample of Starlings from each roost situation is captured and weighed. The results are presented below:

Edge of roost: $\bar{x} = 75.4$ g; $s = 4.14$

$n = 30$ in each case

Centre of roost: $\bar{x} = 83.6$ g; $s = 4.03$

Employing the technique described in Section 9.4, we estimate the 95% confidence interval (C.I.) for each mean, using the *t*-score of 2.045 for a sample size of 30 (= 29 df):

Edge of roost: 95% C.I. $= 75.4 \pm \left(2.045 \times \dfrac{4.14}{\sqrt{30}}\right) = 75.4 \pm 1.55$ g

Center of roost: 95% C.I. $= 83.6 \pm \left(2.045 \times \dfrac{4.03}{\sqrt{30}}\right) = 83.6 \pm 1.50$ g

These results are expressed graphically in Fig. 9.1. The difference between the two sample means is (83.6 − 75.4) = 8.2 g. We record: 'the mean of the sample obtained from the centre is 8.2 g heavier than that obtained from the edge of the roost'. But what are we to infer about the differences between the two populations from which the samples are drawn?

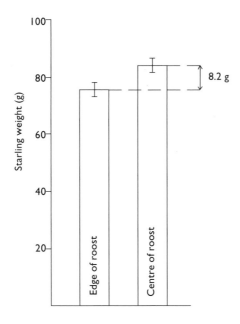

Fig. 9.1 The difference between two sample means.

The problem is, of course, that we only have *estimates* of the two population means. We are 95% confident only that they fall within the respective intervals indicated in Fig. 9.1. It follows that the *difference* between the two sample means is also only an estimate of the difference between the two population means and that we need a method of attaching confidence limits to this estimate.

By similar arguments that were used in the case of the standard error of the mean (Section 9.2) we are about 68% confident that a *difference between two population means* falls within ±1 standard error of a sample mean difference. We now need a formula for working out the **standard error of the difference** between two sample means. It is the square root of the sum of the squared individual sample standard errors:

$$\text{S.E.}_{\text{diff}} = \sqrt{\text{S.E.}^2_1 + \text{S.E.}^2_2}$$

Thus:

$$\text{S.E.}_{\text{diff}} = \sqrt{\frac{(s^2_1)}{(n_1)} + \frac{(s^2_2)}{(n_2)}}$$

For the Starling data (where $n = 30$ in each case)

$$\text{S.E.}_{\text{diff}} = \sqrt{\frac{4.14^2}{30} + \frac{4.03^2}{30}} = 1.055$$

Because the samples are regarded as large (> 30), we multiply the S.E.$_{\text{diff}}$ by the z-score of 1.96 to convert to an approximte 95% confidence interval. We are therefore 95% confident that the true population mean difference $(\mu_1 - \mu_2)$ lies between $(\bar{x}_1 - \bar{x}_2) \pm 1.96$ S.E.$_{\text{diff}}$:

$$
\begin{aligned}
\text{True population mean difference} \quad &= \quad 8.2 \pm (1.96 \times 1.055) \\
&= \quad 8.2 \pm 2.07 \text{ g} \\
&= \quad 10.27 \text{ (upper limit) and } 6.13 \text{ (lower limit).}
\end{aligned}
$$

This result is clearly a good deal more informative than the unprocessed sample mean difference of 8.2 g.

We should note two points here. First, if the larger sample mean is nominated as $\bar{x}_1$ and the smaller as $\bar{x}_2$ then the sample mean difference is conveniently positive. Second, provided that *both* of the samples are quite large (usually more than about 30 observations) it does not matter if they are unequal.

9.7 The difference between the means of two small samples

The rationale for establishing a confidence interval about the difference between the means of two small samples is the same for that of large samples. The formula for estimating the standard error of the difference is a little more complicated, however:

$$
\text{S.E.}_{\text{diff}} \quad = \quad \sqrt{\left[\frac{(n_1 - 1)s^2_1 + (n_2 - 1)s^2_2}{(n_1 + n_2 - 2)} \right] \left[\frac{(n_1 + n_2)}{(n_1 n_2)} \right]}
$$

Although this is a rather cumbersome equation, the terms within it are familiar: they are simply the sample size and standard deviation of each of the two samples being considered.

Example 9.6

Estimate the difference between the means of the two samples of Starling weights for which the sample statistics are given in Example 9.5, if the sample sizes are now 12 and 10 for the edge and centre of roost, respectively.

The difference between the sample means is 8.2 g, as before. Using the equation above (where n_1, s_1 and n_2, s_2 refer to the edge and centre of the roost, repectively):

$$
\begin{aligned}
\text{S.E.}_{\text{diff}} \quad &= \quad \sqrt{\left[\frac{(11 \times 4.14^2) + (9 \times 4.03^2)}{20} \right] \left[\frac{(12 + 10)}{120} \right]} \\[2mm]
&= \quad \sqrt{\frac{(188.54 + 146.17)}{20} \times 0.1833} \\[2mm]
&= \quad \sqrt{3.068} \\[2mm]
&= \quad 1.752
\end{aligned}
$$

Because the samples are small, the use of the z-score of 1.96 is not appropriate; t at $(n_1 + n_2 - 2)$ degrees of freedom is used in its place. For $P = 0.05$, t_{20} is 2.086. The confidence interval is therefore:

$$\text{C.I.} = 8.2 \pm (2.086 \times 1.752)$$
$$= 8.2 \pm 3.655 \text{ g}$$
$$= 11.855 \text{ g (upper limit) and } 4.545 \text{ (lower limit)}$$

This of course represents a considerable extension of the confidence interval estimated for the larger samples of Example 9.5.

9.8 Estimating a proportion

Ornithologists often express the frequency of occurrence of an item in a sample as a **proportion** of the total. For example, an ornithologist catches 80 birds in July, of which 60 are juveniles. The proportion of juveniles in the *sample* is $60/80 = 0.75$ (75%). Provided that adults and juveniles are dispersed independently, this proportion in the sample is an estimate of the actual proportion of juveniles in the *population*. Subsequent samples are all, in turn, independent estimates of the same population proportion but, due to sampling error, are likely to be different in value to each other. In the same way that a set of sample means is clustered around a population mean, so too is a set of sample proportions clustered around a population proportion. The standard deviation of the distribution is similarly called the **standard error**. For practical purposes, a satisfactory estimate of the standard error from a sample proportion is given by:

$$\text{S.E.} = \sqrt{\frac{p(1-p)}{(n-1)}}$$

where p is the proportion of the nominated item and n is the number of all items in the sample.

Example 9.7
Estimate the standard error and 95% confidence interval of the proportion of juvenile birds in a sample which contains 60 juveniles and 20 adults (total number of birds = 80):

$$\text{S.E.} = \sqrt{\frac{0.75(1-0.75)}{(80-1)}} = 0.049$$

The limits of the confidence interval are obtained by multiplying the standard error by 1.96. The interval is:

$$0.75 \pm (1.96 \times 0.049) = 0.75 \pm 0.096$$

That is, 0.846 (84.6%) upper limit and 0.654 (65.4%) lower limit. We are therefore 95% confident that the true proportion of juveniles in the population lies between 0.654 and 0.846 (65.4% and 84.6%).

Two points should be noted here. First, dramatic increases in sample size are required to reduce the confidence interval by an appreciable extent. Thus, increasing the sample size from 80 to 200 (i.e. 250%) in Example 9.7 reduces the standard error from 0.049 to 0.03, a reduction of only 0.019 (40%). Second, the formula for calculating the standard error becomes unreliable when P is less than 0.1 or greater than 0.9.

CHAPTER 10
THE BASIS OF STATISTICAL TESTING

10.1 Introduction

In Section 7.3 we said that statisticians set arbitrary critical thresholds of probability. When an event occurs whose probability is estimated to be below a critical threshold, the outcome is said to be *statistically significant*. The critical values are: $P < 0.05$ (*significant*); $P < 0.01$ (*highly significant*); and $P < 0.001$ (*very highly significant*).

Example 7.7 shows that we can use the properties of the normal curve to estimate probabilities. Thus, an observation obtained randomly from a normally distributed population and having a value exceeding the population mean (μ) plus or minus 1.96 standard deviations occurs in fewer than 1 trial in 20. That is to say, the probability of such an outcome is $P < 0.05$. If a single random observation does exceed the critical value it is regarded as being statistically significant. The procedure for deciding if the outcome of a particular event is significant is called a **statistical test**. We now need to explain in more detail the basis of statistical testing.

10.2 The experimental hypothesis

The formulation and testing of hypotheses is the basis of experimental science. A hypothesis is a proposed explanation for a state of affairs. A hypothesis is tested by experimentation, the outcome of which may provide evidence for the acceptance or rejection of the hypothesis. As an example of an experimental hypothesis we refer to the Starlings roosting communally that we described in Section 9.6. In the statistical sense the roost is divided into two "populations" – one roosting in the centre and the other at the edge, and samples may be drawn from each.

The state of affairs:	Starlings are, on average, heavier at the centre of the roost than at the edge.
Experimental hypothesis:	Socially dominant birds compete successfully for the centre, where microclimate is more favourable.
Possible experiment:	Seek evidence for differences in social behaviour at the centre and edge and try to correlate with microclimate (*Ring. & Migr.* 2,34–37).

If it is found to be windier and colder at the edges of the roost, and it appears that socially "subordinate" birds roost there, then the hypothesis is accepted for the time being. The outcome of the experiment does not *prove* that microclimate is responsible for the discrepancy; it simply fails to provide evidence for doubting it. On the other hand if a difference in microclimate between the edge and centre cannot be detected, the hypothesis is rejected and an alternative proposed.

10.3 The statistical hypothesis

In our example above the description of the *state of affairs* is based on the observation that the mean weight of a sample of starlings drawn from the "population" at the centre is greater than that of a sample drawn from the "population" at the edge. Could there be any reason to doubt the validity of this difference? To examine this suggestion we nominate the mean size of the population at the centre as μ_1 and that from the edge as μ_2. Let us assume for the moment that there is no difference between the means of the two populations, and so $\mu_1 = \mu_2$. (This is actually called a

Null Hypothesis, explained below). If the two population means are indeed identical, we can superimpose the size frequency distribution of one population over that of the other, as we have done in Fig. 10.1. Furthermore, if we assume that there are similar numbers of birds in each population, then the two distributions will have nearly the same height.

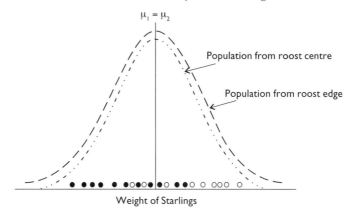

Fig. 10.1 Size frequency distributions of Starlings from two roost situatios. o, random observations from the centre; ●,random observations from the edge.

Now imagine that a small random sample is drawn from each population (that is, a sample from the centre and a sample from the edge of the roost). Although each unit in a sample (a Starling) is drawn randomly it is possible, by chance, that one sample has more than a 'fair share' of larger units. This is due to *sampling error*, described in Section 9.1. If, also by chance, the other sample has more than a 'fair share' of smaller units, then there could be a substantial difference between the means of the two samples, *even though they were drawn from populations with identical means*. This possibility is depicted in Figure 10.1.

The likelihood of such a spurious event arising by chance diminishes rapidly as the sample sizes increase. A statistical test is required to determine the probability that two samples of a given size and with a particular mean difference could have been drawn from two populations with identical means. If the calculated probability is less than the critical threshold, we accept the evidence and conclude that the means of the two samples are statistically significantly different and have therefore been drawn from populations with different means. The basis of a statistical test is a hypothesis that presumes there is *no* difference. This is called a **Null Hypothesis** and is symbolised by H_0. An alternative to a Null Hypothesis is symbolised by H_1. Usually, an alternative hypothesis is simply that there is a difference; for example, $\mu_1 \neq \mu_2$.

By analogy with the experimental hypothesis of Section 10.2 we write:

State of affairs:	The mean weight of a sample of Starlings from the centre of the roost is greater than that of a sample from the edge.
Null Hypothesis H_0:	Although the sample means are different, we assume for the moment that the two samples have been drawn from populations which have identical means, i.e. $\mu_1 = \mu_2$. If H_0 is false then $\mu_1 \neq \mu_2$.
Experiment (statistical test):	A statistical test which calculates the probability that two samples of their particular size and mean difference could have been drawn from two populations with identical means.

If the outcome of the test suggests that the probability of the two samples being drawn from two populations with identical means is too low to be acceptable, we reject H_0 and accept our alternative hypothesis H_1. If this is the case, we may confidently proceed with our fieldwork to establish the cause of the difference, assured that there is little risk that our evidence is based on spurious sampling error.

10.4 Test Statistics

The objective of each of the statistical tests that we shall describe is to produce a single number called a **test statistic**. The important feature of a test statistic is that its probability distribution is known, having been worked out by statisticians; z, whose probability distribution we outline in Section 7.7 is such a statistic. For example, we note in Section 7.7 that the probability (P) of an observation falling outside a z-value of ± 1.96 is less than 0.05; such an observation is therefore considered *significant*. In small samples the same consideration applies to the appropriate value of t.

In each case, a statistical test produces a value for its test statistic, and the task is to determine whether that value exceeds some probability threshold which suggests the rejection of a Null Hypothesis. Fortunately, published tables of the threshold values of each test statistic are readily available and, as we shall show, are easy to use. If a test is performed on *transformed* data, the test statistic is *not* transformed back to the original scale.

10.5 One-tailed and two-tailed tests

In Section 7.8 we distinguish between one-tailed and two-tailed tests. Now we may relate these to the idea of hypothesis testing. In our example in Section 10.3 we establish a Null Hypothesis that two samples are drawn from two populations with identical means. We make no prediction, if H_0 is false, as to which mean is larger than the other – only that they are different:

$$H_0: \mu_1 = \mu_2 \qquad\qquad\qquad H_1: \mu_1 \neq \mu_2$$

If the outcome of the test is such that we are obliged to reject H_0 and accept H_1 we are then entitled to conclude that the sample with the larger mean has been drawn from a population with a larger mean. This conclusion is only reached *after* the test. A test like this which makes no prediction as to which mean is the larger of the two, should they prove to be different, is called a **two-tailed test**.

Sometimes it is possible to predict in advance that if two samples are not drawn from populations with the same mean, then a *nominated* sample has been drawn from a population with a larger mean or, alternatively, has been drawn from a population with a smaller mean. The null hypothesis is the same, but the alternative hypothesis is different. If the population predicted to have the larger mean is nominated μ_1 then:

$$H_0: \mu_1 = \mu_2 \qquad\qquad\qquad H_1: \mu_1 > \mu_2$$

One-tailed tests sometimes arise in taxonomy. For example, if a sample does not represent population A, then it must represent population B, which is known to be smaller. Because a one-tailed test is less stringent than a two-tailed test, considerable caution should be exercised before using it. Moreover, 'significant' results can be obtained with smaller sample sizes. Observers who persuade themselves that a test is one-tailed in order to obtain a "result" are *definitely cheating!*

In summary, a one-tailed test should be used only when there is an *advance* reason to predict a directional influence in the data; moreover, the decision to use a one-tailed test *must* be made before the data are analysed. If there is doubt, use a two-tailed test; an outcome which is significant in a two-tailed test is also significant in a one-tailed test. There is no difference between the execution of a one-tailed test and a two-tailed test.

10.6 Hypothesis testing and the normal curve

We may now relate the principles outlined in this chapter to the statistical tests undertaken in Chapter 7. Turn again to Example 7.7. There we ask if it is likely that a single observation of 16.4 mm is drawn randomly from a population for which $\mu = 15.2$ mm and $\sigma = 0.55$ mm. Our null hypothesis H_0 is that the observation *is* drawn from such a population (H_1, that it is not). The test involves the computation of a test statistic z. The computed value of z (2.18) exceeds 1.96 which is the value corresponding to the critical probability threshold of $P = 0.05$. We therefore reject H_0, accept H_1 and conclude that the observation is probably not drawn from the population.

In a one-tailed test, the $P = 0.05$ significance level lies at the z score of 1.65 on whichever side of the mean that is subject to test. Note that it is less stringent than the value of 1.96 of a two-tailed test. Remember that if estimates of μ and σ are based on small samples, then the z should be replaced by t at the appropriate number of degrees of freedom obtained from the table in Appendix 1.

Figure 10.2 illustrates how the normal curve is used in hypothesis testing.

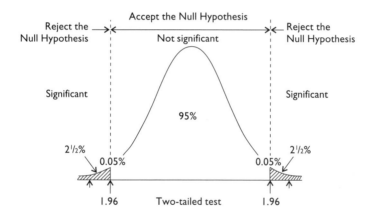

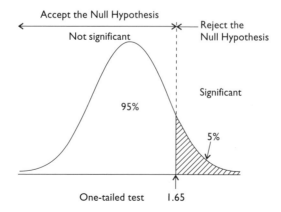

Fig. 10.2 Hypothesis testing and the normal curve: (a) two-tailed test; (b) one-tailed test.

10.7 Type I and type 2 errors

When the threshold for rejection of H_0 is set at $P = 0.05$, an investigator is said to *accept the 0.05 (or 5%) level of significance*. This means that in tests where the computed value of the test statistic is about equal to the critical value, the decision to reject H_0 is probably correct 19 times out of 20 or 95 times out of 100. We expect in the long run that about 5 cases in 100 there is a risk of rejecting H_0 when it is true. When H_0 is rejected and it is actually true, we refer to a **type 1 error** having been committed. How can the risk of a type 1 error be reduced? Simply by setting the level of acceptance at a more rigorous standard, for example, at the 1 in 100 times level of significance ($P = 0.01$). It will be appreciated, however, that the analyst faces a 'swings and roundabouts' situation. The opposite of the case just outlined is referred to as a **type 2 error**, that is not rejecting the null hypothesis when, in fact, it should be rejected.

> *Thus, as the likelihood of making a type 1 error is reduced, the likelihood of making a type 2 error is increased.*

In most statistical analyses the aim is usually to limit the probability of committing type 1 errors, thus erring on the side of caution. In practice, the calculated value of a test statistic often exceeds the tabulated critical value at $P = 0.05$ in which we reject H_0 at $P < 0.05$ and the risk of error is accordingly reduced.

10.8 Parametric and non-parametric statistics: some further observations

In Section 2.7 brief reference was made to the distinctions between parametric and non-parametric statistics. It is noted that the conditions relating to the use of parametric tests are more rigorous than those which apply to non-parametric tests. These distinctions are now elaborated upon in Table 10.1.

Table 10.1 Distinctions between non-parametric and parametric tests

Non-parametric tests	Parametric tests
May be used with actual observations, or with observations converted to ranks	Are used only with actual observations
May be used with observations on nominal, ordinal and interval scales	Generally restricted to observations on interval scales
Compare medians	Compare means and variances
Do not require data to be normally distributed or to have similar variances i.e. they are 'distribution free'	Require data to be normally distributed and to have similar variances
Are suitable for data which are counts	Counts must usually be transformed
Are suitable for derived data, e.g. proportions, indices	Derived data may first have to be transformed

Ornithologists may wish to test nominal and ordinal level data in which case non-parametric tests are often very suitable. Non-parametric tests usually employ simple formulae and avoid the rather tedious computations of variances, sums of squares, etc. that are often required in parametric tests. Furthermore, they require no data transformation. On the other hand, because non-parametric tests do not use *all* the data (that is **ranks** rather than actual observations are used) and are more 'permissive', they may be less powerful than a corresponding parametric test.

10.9 The power of a test

Statisticians sometimes refer to the **power of a test**. It is a measure of the likelihood of a test reaching the correct conclusion, i.e. accepting H_0 when it should be accepted. Non-parametric tests are generally regarded as being less powerful than parametric equivalents, because they are less rigorous in their conditions of use. It must be emphasised, however, that parametric tests are only more powerful than non-parametric when the assumptions governing their use hold true. If an assumption does not hold true (e.g. data do not fit a normal distribution), it is not just a question of a reduction of power; the whole validity of the test is destroyed and there may be the risk of considerable error. There is one safe rule:

If there is doubt as to whether a particular set of data satisfies the assumptions made in the use of a parametric test, then a non-parametric alternative should be used.

CHAPTER 11
ANALYSING FREQUENCIES

11.1 The chi-square test

Ornithologists spend a good deal of their time counting and classifying things on nominal scales such as species, age-class and habitat. Statistical techniques which analyse frequencies are therefore especially useful. The classical method of analysing frequencies is the chi-square test ("chi" is the Greek letter χ and is pronounced "ky" as in "sky"). This involves computing a test statistic (χ^2) which is compared with the probability distribution of the statistic printed in tables. Commonly, such tables of the distribution of χ^2 are restricted to the critical values at the significance levels we are interested in. In Appendix 2 we give critical values at $P = 0.05$ and $P = 0.01$ (the 5% and 1% levels) for 1 to 30 df. Between 30 and 100 df, the critical values are estimated by interpolation, but the need to do this arises infrequently.

   Chi-square tests are variously referred to as tests for **homogeneity, randomness, association, independence** and **goodness of fit**. The array is not as alarming as it might seem at first sight. The precise applications will become clear as you study the examples. In each application the underlying principle is the same. The frequencies we *observe* are compared to those we *expect* on the basis of some null hypothesis. If the discrepancy between observed and expected frequencies is great, then the calculated value of the test statistic will exceed the critical value at the appropriate number of degrees of freedom. We are then obliged to reject the null hypothesis in favour of some alternative.

   The mastery of the method lies not so much in the calculation of the test statistic itself but in the calculation of the expected frequencies. We have already shown an example of how expected frequencies may be generated (see Example 7.6).

11.2 Calculating the χ^2 test statistic

The simplest arithmetical comparison that can be made between an observed frequency and an expected frequency is the difference between them. In the test, the difference is squared and divided by the expected frequency. Thus:

$$\chi^2 = \frac{(O-E)^2}{E}$$

where O is an observed frequency and E is an expected frequency.

   In practice, a series of observed frequencies is compared with corresponding expected frequencies resulting in several components of χ^2. The general formula for χ^2 is therefore:

$$\chi^2 = \sum \frac{(O-E)^2}{E}$$

   When a series of chi-square values are summed we have to take into account the number of degrees of freedom. How to work out the df is described in the examples which follow.

11.3 An example of the use of chi-square

Example 11.1

Let us suppose that a bird ringer wishes to test the effectiveness of three different traps for catching birds. A trap of each design is set up and their positions interchanged periodically to eliminate the effects of other variables. The numbers of birds captured in each trap design over the study period are recorded below.

Design A	Design B	Design C	Total
10	27	15	52

These are, of course, the **observed frequencies**. We are clearly dealing with nominal categories: a trap is either design A or B or C.

The question we are interested in is, "Does the observation that more birds were caught in trap design B really reflect a genuine difference, or could the discrepancy be due to chance scatter or **sampling error**?" Another way of asking the question is, "Is the distribution of frequencies between the traps homogeneous (i.e. evenly spread)?" The test really amounts to a test of "even-ness" or **homogeneity**. The statistical hypotheses are therefore:

H_0: The observed frequencies are homogeneous and any departure can be accounted for by chance scatter or sampling error;

H_1: The observed frequencies depart from those expected of a homogeneous (even) distribution by an amount that cannot be explained by sampling error.

The next step is to ask what frequencies would have been *expected* if H_0 is indeed true. If the frequencies reflect a homogeneous distribution, we would expect the 52 birds to be distributed equally between all three designs. That is $52/3 = 17.33$ (or thereabouts) in each design. We can now write out the observed and expected frequencies together and calculate χ^2.

Design	Observed frequency	Expected frequency	Obs. – Exp.	(Obs. – Exp.)^2/Exp.
A	10	17.33	−7.33	3.10
B	27	17.33	9.67	5.40
C	15	17.33	−2.33	0.313
				$\chi^2 = 8.813$

Having calculated the test statistic, the next step is to evaluate the number of degrees of freedom (df). The rule is simple. It is the number of categories, *a*, less one. In this example there are therefore 2 df. We now have to determine whether the calculated value of 8.813 at 2 df exceeds the critical value at a chosen level of significance. This is decided by reference to Appendix 2. Turn to Appendix 2 and run down the left column until the number of degrees of freedom (2) is reached. Looking across the table at that point we find two values: 5.99 under the 0.05 level of significance, and 9.21 under the 0.01. Our value of 8.813 is bigger than the first but smaller than the second. Thus, the value of the test statistic exceeds the threshold at 0.05 (5%) but not at 0.01 (1%). We conclude that the discrepancy between the observed and expected frequencies is statistically 'significant' but not 'highly significant'. In plain language this means that if in fact there is no underlying difference between the effectiveness of the traps, we should expect a variation of this magnitude *by chance* in fewer than 5 trials out of 100, but in more than one trial in 100. These odds are long enough for us to presume that there is a difference. The result could be recorded: "The trap of design B was shown in a trial to be significantly more effective in catching birds than the other two designs tested; $\chi^2_2 = 8.813$, $P < 0.05$". The subscript 2 below the χ^2 is a shorthand way of showing the two degrees of freedom.

11.4 One degree of freedom – Yates' correction

When there are only two categories in a distribution there is only one degree of freedom. In this case, the calculated value of the test statistic is too high unless we make a correction called **Yates' Correction for Continuity**. This involves subtracting 0.5 from the numerator of each component of the chi-square formula before squaring. The subtraction is made from the *absolute value* of the difference $(O - E)$, that is, any minus sign is ignored. This is written as $(|O - E| - 0.5)^2$ where the vertical bars on each side of $(O - E)$ mean *absolute value*.

Example 11.2

A ringer captures 16 Reed Buntings entering a communal roost; 12 are males and 4 are females. The question arises: "Are there significantly more males in the catch than females?"; alternatively, "Does a ratio of 12:4 constitute a significant departure from a sex ratio of unity?". In either case, the test is again one of homogeneity, seeking departure from a null hypothesis of an even (50:50) distribution between the two sex categories.

If the null hypothesis that the sex ratio is unity is true, we should expect 8 males and 8 females in the sample of 16. Because there is 1 df we apply Yates' correction:

- *without* the correction, the value of the test statistic is $[(12 - 8)^2/8] + [(4 - 8)^2/8] = (2.0 + 2.0) = 4.0$. The value exceeds the critical value of 3.84 at 1 df and is significant at $P = 0.05$.

- *with* the correction it is $[(|12 - 8| - 0.5)^2/8] + [(|4 - 8| - 0.5)^2/8 = (1.531 + 1.531) = 3.062$. This is *less* than the tabulated critical value of 3.84 at $P = 0.05$.

We conclude that there is *no* significant departure of the sex ratio from 1:1. Notice how in this particular example the application of the correction alters the outcome of the test.

11.5 Tests for association – the contingency table

In the previous examples of chi-square analysis, observed frequencies are distributed between categories in *one row*. When this is the case we refer to a **one-way classification**. Sometimes, however, two nominal level observations are obtained from a sampling unit. Thus, we may record an individual according to its age class and sex; its species and habitat; and so on. In such cases, frequencies are arranged in *two or more rows* and we refer to a **two-way classification**. Tables of these data are called contingency tables. They allow the investigation of *association* between variables. The simplest type of contingency table is one which has only two nominal categories of each variable. It is called a **2 x 2 table**. An example follows.

Example 11.3

An ornithologist surveys two *Common Bird Census* plots, one in a coniferous wood and one in a deciduous wood. The number of territories of Coal Tits and Blue Tits are presented in a contingency table (Table 11.1). The cells in the table are conventionally labelled *a, b, c, d* and row totals, column totals and grand total are included.

Table 11.1 Numbers of Blue Tit and Coal Tit territories

	Blue Tit	Coal Tit	Totals
Deciduous wood	a 14	b 6	a + b 20
Coniferous wood	c 22	d 46	c + d 68
Totals	a + c 36	b + d 52	a+b+c+d 88

Inspecting the distribution of frequencies in the table we see that there are more Blue Tits than Coal Tits in the deciduous habitat, but more Coal Tits than Blue Tits in the coniferous. If a statistical test were to show that the proportional difference could not be accounted for by sampling error, then we would say they are significantly different. This would mean that there is an *interaction* between the two variables *species* and *habitat*. The interaction could be a positive association or a negative association. It appears that Blue Tits are positively associated with the deciduous habitat and negatively associated with coniferous. If, however, a test were to show there is *no* significant difference between the proportions, then we would conclude that there is no interaction and the variables are independent. The test may therefore be employed as a *test for association* or *a test for independence*, according to your point of view.

In order to apply the chi-square test we need to calculate the expected frequency in each cell. This task is complicated by the fact we are in effect dealing with two null hypotheses:

(i)   the ratios of the frequencies in both vertical columns do not depart from the overall vertical ratio (i.e. 14:22 and 6:46 do not depart from 20:68);

(ii)   the ratios of the frequencies in both horizontal rows do not depart from the overall horizontal ratio (i.e. 14:6 and 22:46 do not depart from 36:52).

To calculate the expected frequency for any single cell we apply the formula: column total x row total $\div$ grand total. Thus, the expected frequency of Blue Tits in *deciduous* is $[(a + c) \times (a + b)] \div (a + b + c + d) = 36 \times 20 \div 88 = 8.182$. The table may now be written out (in abbreviated form) showing the observed and expected frequencies in each cell.

Cell a $O = 14$ $E = 36 \times 20 \div 88 = 8.182$	**Cell b** $O = 6$ $E = 52 \times 20 \div 88 = 11.818$
Cell c $O = 22$ $E = 36 \times 68 \div 88 = 27.818$	**Cell d** $O = 46$ $E = 52 \times 68 \div 88 = 40.182$

(Check the sum of the expected frequencies equals the sum of the observed frequencies, $a + b + c + d = 8.182 + 11.818 + 27.818 + 40.182 = 88$).

With the four observed frequencies and their respective expected frequencies, the individual components of the test statistic may now be calculated. Before we do that, it is advisable to determine the number of degrees of freedom. The rule for determining the degrees of freedom in a contingency table is:

Degrees of freedom = number of columns (c) minus 1, multiplied by the number of rows (r) minus 1 that is: $(c-1)(r-1)$.

In our example this is $(2-1)(2-1) = 1$.

A 2 x 2 contingency table, therefore, has only one degree of freedom. With the conditions described in the previous section in mind, we must apply Yates' correction to each cell in the table. Readers may note that a 2 x 2 table is the only one in which the correction needs to be applied. Taking the four cells respectively, the test statistic is the sum of:

Cell a	Cell b
$(\mid 14-8.182\mid - 0.5)^2/8.182$	$(\mid 6-11.818\mid - 0.5)^2/11.818$
$= 3.457$	$= 2.393$
Cell c	Cell d
$(\mid 22-27.818\mid - 0.5)^2/27.818$	$(\mid 46-40.182\mid - 0.5)^2/40.182$
$= 1.017$	$= 0.704$

$$\chi^2 = 3.457 + 2.393 + 1.017 + 0.704 = 7.571.$$

Consulting Appendix 2 we find that the calculated value of 7.571 at 1 df exceeds the tabulated critical value at both 5% and 1% levels of significance. There is therefore a statistically highly significant association between the variables. Blue Tits are associated with deciduous habitats and Coal Tits with coniferous.

11.6 The r x c contingency table

Where there are more than two nominal categories in a two-way classification a contingency table has several rows and columns in it. If there are r rows and c columns there are r x c cells in the table. The procedure for working out the expected frequencies and calculating the test statistic is similar to that of a 2 x 2 table except that Yates' correction is not applied.

Example 11.4

An ornithologist studying a Starling roost suspects that the dispersion of birds within the roost is not random, but there is some degree of "social ordering". Recognisable categories within the Starling population are adult males, immature males, adult females and immature females. Departure from a random dispersion in the roost, it is hypothesised, is reflected in an association between one or more of the population categories and certain parts of the roost. In order to test this hypothesis, Starlings are captured in three different parts of the roost where it is anticipated there are differences. The frequencies obtained are shown in Table 11.2.

Table 11.2 Frequency of Starling categories

	Population category				
	Adult male	Immature male	Adult female	Immature female	Totals
Location 1	22	10	20	5	57
Location 2	10	13	6	9	38
Location 3	10	7	6	15	38
Totals	**42**	**30**	**32**	**29**	**133**

All categories in both directions (i.e. population category and location) are truly nominal and so chi-square contingency analysis is appropriate. The steps in the analysis are explained below.

(i) Calculate the 12 individual frequencies that would be expected if H_0 is true, that is, there is no association between any of the categories. By way of example, the expected frequency of adult male in *location 1* is 42 x 57 ÷ 133 = 18.0 (column total x row total ÷ grand total).

(ii) Calculate all the individual components of χ^2 from $(O - E)^2/E$. The results of the procedure thus far obtained are tabulated in Table 11.3.

Table 11.3 Calculation of expected frequencies

	Population category			
	Adult male	Immature male	Adult female	Immature female
Location 1				
O:	22	10	20	5
E:	18	12.9	13.7	12.4
$(O - E)^2/E$	0.89	0.65	2.90	4.42
Location 2				
O:	10	13	6	9
E:	12	8.57	9.14	8.29
$(O - E)^2/E$	0.33	2.29	1.08	0.06
Location 3				
O:	10	7	6	15
E:	12	8.57	9.14	8.29
$(O - E)^2/E$	0.33	0.288	1.08	5.43

(iii) Sum all the individual values of $(O - E)^2/E$. The result is 19.75. This is the test statistic.

(iv) Determine the degrees of freedom from $(c - 1)(r - 1)$. This is 3 x 2 = 6.

(v) Consult Appendix 2 at this number of degrees of freedom and decide if the test statistic calculated in step (iii) exceeds the critical value at $P = 0.05$ or $P = 0.01$. The critical value at $P = 0.01$ is 16.81. The calculated value exceeds this.

(vi) Express the result in the form: "There is a statistically highly significant association between certain population categories and particular roosting sites; $\chi^2_6 = 19.75$, $P < 0.01$".

In order to decide which population categories are associated with which particular roosting location, we examine the individual components of χ^2 in the Table 11.3. Scanning the table, we find the largest individual values are 4.42 (immature females in location 1) and 5.43 (immature females in location 3). Looking at those cells of the table more closely, we note that there are more immature females than expected in location 3, and fewer than expected in location 1. We conclude that there is an association between certain population categories and different parts of the roost. In particular, immature females seem to prefer location 3 and avoid location 1.

11.7 The G-test

The **G-test** is an alternative to the chi-square test for analysing frequencies. The two methods are interchangeable; if a chi-square test is appropriate then so too is a G-test and the assumptions in each are the same. Moreover, the outcome of the G-test is a test statistic G which is compared with the distribution of chi-square in the same tables as the chi-square test. Why then do we need a second test that serves exactly the same purpose?

First, the G-test is easier to execute with a desk-top hand calculator, especially with contingency tables. Second, mathematicians believe that the G-test has theoretical advantages in advanced applications which are beyond the scope of this Guide. Despite these advantages, however, it seems that 'old habits die hard'; chi-square is still the most widely used method of analysing frequencies in current ecological journals.

In one-way classifications and in 2 x 2 contingency analyses a correction factor (**Williams' correction**) is normally applied to the calculated value of G. The adjusted value of G is accordingly denoted G_{adj}.

11.8 The G-test as a test for homogeneity

In order to emphasise the similarity between the G-test and the chi-square test, we use the data employed in Example 11.1 to illustrate the use of the G-test as a test for homogeneity in a one-way classification.

Example 11.5

The observed and expected frequencies of birds caught in the three trap designs were:

O:	10	27	15
E:	17.33	17.33	17.33

The formula for G is:

$$G = 2 \times \overset{a}{\Sigma} \, O \ln \frac{O}{E}$$

where O and E are observed and expected frequencies, respectively; $\overset{a}{\Sigma}$ means the sum of the products $O\ln(O/E)$ for all a categories; and 'ln' means natural logarithm. The stepwise procedure is as follows:

Step 1

For each category multiply O by $\ln(O/E)$ and add up the total. Thus:

$$10 \times \ln \left(\frac{10}{17.33} \right) + 27 \times \ln \left(\frac{27}{17.33} \right) + 15 \times \ln \left(\frac{15}{17.33} \right)$$

$$= -5.50 + 11.97 + (-2.166) = 4.30$$

Step 2

Double this number:

$$4.30 \times 2 = 8.6$$

This is the test statistic, G.

Step 3

Divide G by the correction factor which is applied irrespective of the number of degrees of freedom.

$$\text{Correction factor} = 1 + (a^2 - 1)/6n\nu$$

where a is the number of categories, n is the total number of observed frequencies and ν is the degrees of freedom $(a - 1)$. Thus:

$$
\begin{aligned}
\text{Correction factor} &= 1 + (3^2 - 1)/(6 \times 52 \times 2) \\
&= 1 + (8/624) \\
&= 1.0128 \\
G_{adj} &= G/\text{correction factor} = 8.6/1.0128 = 8.491
\end{aligned}
$$

Step 4

Compare the value of G_{adj} against the chi-square distribution (Appendix 2) for $(a - 1) = 3$ df.

The calculated value of G_{adj} is very similar to the value of 8.813 obtained in the chi-square test in Example 11.1.

There is no difference in the execution of the G-test when there is only one degree of freedom. Readers may apply the G-test to the data in Example 11.2 and confirm that $G = 4.186$, the correction factor is 1.031 and G_{adj} is 4.06.

11.9 Applying the G-test to a 2 x 2 contingency table

In using the G-test for analysing 2 x 2 and r x c contingency tables we do not have to distinguish between observed and expected frequencies. We symbolise the observed frequencies therefore simply as f. Observed frequencies are multiplied by the natural logarithms of themselves and the products summed $(\Sigma f.\ln f)$. This operation is easily accomplished with a scientific calculator. The procedure for using the G-test with a 2 x 2 contingency table is shown in Example 11.7.

Example 11.7

Bridled and unbridled Guillemots are counted in sampling units on Fair Isle and St. Kilda (a sampling unit being defined as an area of cliff face that can be seen clearly and safely and in which birds can be counted accurately). The Fair Isle sample comprised 31 bridled and 89 unbridled birds, whilst that from St. Kilda comprised 35 bridled and 115 unbridled. That is, 26% bridled on Fair Isle and 23% on St. Kilda. Is the difference in proportional frequency statistically significant?

Step 1

Display the observed frequencies in a contingency table showing row, column and grand totals.

	Bridled	Unbridled	Totals
Fair Isle	a 31	b 89	a + b 120
St. Kilda	c 35	d 115	c + d 150
Totals	a + c 66	b + d 204	a + b + c + d 270

Step 2

Calculate $\Sigma f.\ln f$ for all observed frequencies:

$$(31.\ln31) + (89.\ln89) + (35.\ln35) + (115.\ln115) = 1176.047$$

Step 3
Calculate $\Sigma f.\ln f$ for the grand total $(a + b + c + d)$:

$$270.\ln270 = 1511.574$$

Step 4
Calculate $\Sigma f.\ln f$ for all row and column totals:

$$(120.\ln120) + (150.\ln150) + (66.\ln66) + (204.\ln204) = 2687.508$$

Step 5
Add the numbers obtained in Steps 2 and 3 and subtract the number obtained in **Step 4**:

$$\text{Step 2} + \text{Step 3} - \text{Step 4}$$

$$1176.047 + 1511.574 - 2687.508 = 0.113$$

Step 6
Double this number to give G

$$G = 0.113 \times 2 = 0.226$$

Step 7
Calculate Williams' correction factor from:

$$\text{Correction factor} = 1 + \frac{\left[\left(\dfrac{n}{a+b}\right) + \left(\dfrac{n}{c+d}\right) - 1\right]\left[\left(\dfrac{n}{a+c}\right) + \left(\dfrac{n}{b+d}\right) - 1\right]}{6n}$$

$$= 1 + \frac{\left[\left(\dfrac{270}{120}\right) + \left(\dfrac{270}{150}\right) - 1\right]\left[\left(\dfrac{270}{66}\right) + \left(\dfrac{270}{204}\right) - 1\right]}{6 \times 270}$$

$$= 1 + \frac{13.46}{1620}$$

$$= 1 + 0.0083 = 1.0083$$

Step 8
Divide G by the correction factor to obtain G_{adj}:

$$G_{adj} = 0.226/1.0083 = 0.224$$

Step 9
Compare the value of G_{adj} with chi-square in Appendix 2 at $(r-1)(c-1) = 1$ df. The value is well below the critical value of 3.84 at $P = 0.05$. We accept the null hypothesis and conclude that there is no statistically significant difference between the proportions of bridled and unbridled Guillemots at Fair Isle and St. Kilda. This does not necessarily mean that a real difference does not exist, merely that the data to hand are insufficient to show it. Larger counts might give a different result.

11.10 Applying the G-test to an r x c contingency table

The procedure for applying the G-test to an $r \times c$ contingency table is exactly the same as for a 2×2 table. However, the correction factor is not usually applied because it is so small that it may be ignored.

Example 11.8

In a study of lowland grassland as a habitat for breeding waders, 123 grassland sites were classified according to their conservation status as nature reserves (26), sites of special scientific interest (SSSI 25), or unprotected (72). A survey of breeding Snipe was conducted in 1982 and then repeated in 1989 to discover if numbers had declined, were stable or had increased in the three categories. The results are given in Table 11.4 (from *Bird Study*, 1992, 29, p.172).

Inspection of the data suggests that the smallest proportional decline was in reserves, but is this statistically significant? Because all categories in the rows and columns are truly nominal, the G-test is appropriate for this problem. The steps in the tests are as follows:

Table 11.4 Snipe in grassland habitats

	Number of sites with			
	Decline	Stable	Increase	Total
Reserves	4	14	8	26
SSSI	11	6	8	25
Unprotected	31	13	28	72
Total	46	33	44	123

Step 1

Display the observed frequencies in a contingency table showing row, column and grand totals (this has been done in the table above).

Step 2

Calculate $\Sigma f.\ln f$ for all observed frequencies:

$(4.\ln 4) + (14.\ln 14) + (8.\ln 8) + (11.\ln 11) + (6.\ln 6) + (8.\ln 8) + (31.\ln 31) + (13.\ln 13) + (28.\ln 28)$
$= 5.545 + 36.947 + 16.636 + 26.377 + 10.751 + 16.636 + 106.454 + 33.344 + 93.302$
$= 345.992$

Step 3

Calculate $f.\ln f$ for the grand total:

$123.\ln 123 = 591.899$

Step 4

Calculate $\Sigma f.\ln f$ for all row and column totals:

$(26.\ln 26) + (25.\ln 25) + (72.\ln 72) + (46.\ln 46) + (33.\ln 33) + (44.\ln 44)$
$= 84.711 + 80.472 + 307.920 + 176.118 + 115.385 + 166.504$
$= 931.110$

Step 5

Calculate: Step 2 + Step 3 − Step 4

$$345.992 + 591.899 − 931.11 = 6.781$$

Step 6

Double this to give G

$$6.781 \times 2 = 13.562$$

Step 7

No correction is applied. Compare G with the chi-square distribution in Appendix 2 at $(r − 1)$ $(c − 1) = 2 \times 2 = 4$ df. The calculated value of G is larger than the critical value of 13.28 at $P = 0.01$. We therefore conclude that there are indeed proportional differences between the habitat categories, and Snipe have declined less in protected sites.

11.11 Restrictions and cautions

1. All versions of the chi-square test compare the agreement between a set of observed frequencies and those expected if some null hypothesis is true. G-tests are similar, but note that in contingency tables the intermediate calculation of expected frequencies is not required in the G-test.

2. As objects are counted they should be assigned to nominal categories. Unambiguous intervals on a continuous scale may be regarded as nominal for the application of the tests, however. Frequencies obtained from contiguous time bands or contiguous longitudinal or altitudinal zones, for example, are acceptable nominal categories.

3. The application of the tests requires that the samples are random and the objects being counted are independent. Thus, in Example 11.7, it is assumed that the bridled Guillemots are dispersed independently through the colony, and are not attracted to each other. If this were the case, the objects being counted would not then be independent and the application of a chi-square or G-test would be invalid.

4. In the chi-square tests, the sample size (that is, the grand total of observed frequencies) should be such that all *expected* frequencies exceed 5. In marginal cases this can sometimes be achieved by collapsing cells and aggregating the respective observed frequencies and expected frequencies. Some flexibility in interpreting this rule is allowed. Most statisticians would not object to some of the expected frequencies being below 5, provided that no more than one-fifth of the total number of expected frequencies are below 5, and none are below 1.

5. All versions of the tests require that observed frequencies are, indeed, *actually observed.* They must not be estimates or derived variables. For example, in Example 11.7 the bridled Guillemots could be expressed as 26% on Fair Isle and 23% on St. Kilda. Without the actual number of birds counted, however, we are unable to test the difference between the proportions.

6. Apply Yates' correction in the chi-square test when there is only one degree of freedom. In the G-test apply Williams' correction in all one-way classification tests and in 2 x 2 contingency tables. It may be ignored in larger contingency tables.

7. The problem might arise whether to chose the chi-square or the G-test. They are totally interchangeable and if one is applicable, then so too is the other. The underlying assumptions of each are the same. In analysing contingency tables the G-test is quicker to execute using a

scientific calculator because the expected frequencies are not separately calculated, as in the chi-square test. On the other hand, a direct comparison of the observed with the expected frequencies can be helpful in pin-pointing relationships (see Example 11.4). Which test you are to employ is decided at the outset of a project and then used throughout. You should definitely *not* try out each test to see if one gives a slightly more 'favourable' result – that is cheating! In the ornithological literature the chi-square test is still the favourite but the G-test appears to be gaining in popularity as its flexibility is recognised. For further information on the application of the G-test we suggest you consult Sokal and Rohlf (1981).

CHAPTER 12
MEASURING CORRELATIONS

12.1 The meaning of correlation

Many variables in nature are related; examples from ornithology include the weight of a growing chick and its age; the number of nests in a wood and its area. Relationships or associations between variables such as these are referred to as correlations. Correlations are measured on ordinal or interval scales.

When an increase in one variable is accompanied by an increase in another, the correlation is said to be **positive** or **direct**. Number of nests and area of woodland are positively correlated. When an increase in one variable is accompanied by a decrease in another, the correlation is said to be **negative** or **inverse**. The weight of body fat of a migrating bird and the distance flown since its last feed are negatively correlated.

The fact that the variables are associated or correlated does not necessarily mean that one causes the other – the two may be independently related to a third (perhaps unidentified) factor. Thus, wing length and tail length of individuals in a population may be correlated but one cannot be said to cause the other – both are undoubtedly related to some underlying genetic factor. In common usage, the word 'correlation' describes any type of relationship between objects and events. In statistics however, correlation has a precise meaning; it refers to a quantitative relationship between two variables measured on ordinal or interval scales.

12.2 Investigating correlation

Bivariate observations of variables measured on ordinal or interval scales can be displayed as a scattergram (Figs. 4.9 and 12.1). Just as a simple dot-diagram gives a useful indication of whether a sample of observations is roughly symmetrically distributed about a mean and the extent of the variability, a scattergram gives an impression of correlation. Figure 12.1(a) shows a clear case of a positive correlation, whilst Fig. 12.1(b) shows an equally clear case of a negative correlation. Fig. 12.1(c) shows no correlation, but what about Fig. 12.1(d)? It is not so easy to be certain about this.

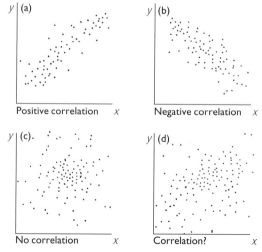

Fig. 12.1 Scattergrams of bivariate data

Subjective examination of a scattergram must be replaced by a statistical technique which is more precise and objective. The statistic that provides an index of the degree to which two variables are related is called the **correlation coefficient**. The statistic is calculated from sample data and is the estimator of the population parameter.

The numerical value of the correlation coefficient, r, falls between two extreme values: $+1$ (for perfect positive correlation) and -1 (for perfect negative correlation). A perfect correlation exists when all the points in a scattergram fall on a perfectly straight line, as indicated in Fig. 12.2. Perfect or near perfect correlations (positive or negative) are virtually non-existent in biological situations; they tend to be the privilege of the physicist! An example of a perfect correlation is described in Section 13.4. A correlation coefficient of 0, or near 0, indicates lack of correlation.

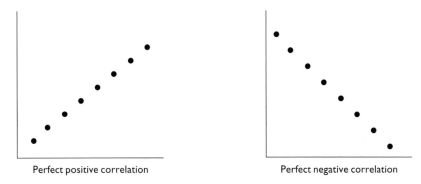

Perfect positive correlation   Perfect negative correlation

Fig. 12.2 Perfect correlations

Correlation coefficients can be calculated by both parametric and non-parametric methods. A parametric coefficient is the **Product Moment Correlation Coefficient** (sometimes called the **Pearson Correlation Coefficient**). It is used only for interval scale observations and is subject to more stringent conditions than non-parametric alternatives. Of the various non-parametric coefficients, the **Spearman Rank Correlation Coefficient** is among the most widely used and is the one we illustrate. It is appropriate to observations based on ordinal as well as interval scales.

12.3 The strength and significance of a correlation

All values of a correlation coefficient fall on a scale with limits -1 to $+1$. The closer the value is to -1 or $+1$, the greater is the strength of the correlation, whilst the closer it is to zero the weaker it is. As a rough and ready guide to the meaning of the coefficient Table 12.1 offers a descriptive interpretation.

Table 12.1 The strength of a correlation

Value of coefficient r (positive or negative)	Meaning
0.00 to 0.19	A very weak correlation
0.20 to 0.39	A weak correlation
0.40 to 0.69	A modest correlation
0.70 to 0.89	A strong correlation
0.90 to 1.00	A very strong correlation

Sometimes an apparently strong correlation may be regarded as *not statistically significant* whilst a weak correlation may be *statistically highly significant*. We must resolve this apparent paradox. Look again a Fig. 12.1(c) which we say shows no correlation. Imagine that the points in the scattergram represent a population of observations from which individual points can be drawn at random. The first two points drawn are bound to be in perfect alignment because any two points can be connected by a straight line. We should not be tempted however into thinking in terms of correlation! It is not improbable that a third point, drawn randomly, could be in rough alignment with the first two, giving an impression of correlation. The more points that are drawn the less and less likely it becomes that a chance impression of correlation can be maintained. When *all* the points in the population have been drawn, we can measure the absolute correlation coefficient parameter ρ (rho). Any sample drawn from the population estimates ρ. Large samples give reliable estimates and small samples give less reliable estimates. If the value of ρ is low, that is, a weak correlation in the population, large samples will give good estimates and are *statistically significant*. On the other hand, if ρ should be large, a small sample yields a poor estimate which may not be statistically significant. Thus, a coefficient of value 0.2 is not significant in a sample size of 40 units; it is, however, highly significant in a sample of 175 units. The point to bear in mind is that you cannot 'improve' or 'strengthen' a correlation by increasing the sample size; but you will reduce the likelihood of a spurious correlation arising by chance sampling error.

12.4 The Product Moment Correlation Coefficient

The Product Moment Correlation Coefficient is a parametric statistic which is appropriate when observations are measured on interval scales from randomly sampled units. It is assumed that *both* variables are approximately normally distributed, that is to say *bivariate normal*. This can be checked from a scattergram of the data. The broad outline of points in a scattergram of bivariate normal data is roughly circular or elliptical. The circle becomes drawn out into an ellipse as *r* increases in value (Fig. 12.3).

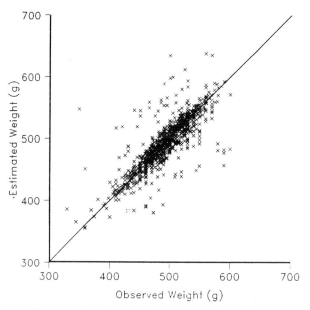

Fig. 12.3 Scattergram of observed weights of albatross eggs and weights estimated from measurements showing a very strong correlation between bivariate normal variables (from Ibis, 1992, 134, p.220).

Example 12.1

Members of a Wader Study Group catch an unexpectedly large number of Dunlin in a single catch. Normally painstaking in the recording of every useful measurement, they now face the choice of either reducing the number of measurements recorded from each bird or releasing some of the birds unprocessed to ensure that all birds are released within an acceptable time. If is established that two of the variables usually measured (say weight and bill length) are strongly correlated, it might be considered not too detrimental to the study to forego one of the measurements (say, the most time-consuming one) in the knowledge that the other variable will still provide information needed to estimate, for example, size variability within the population.

The weights and bill-lengths of 10 Dunlin are presented in Table 12.2, together with columns for x^2, y^2 and xy. To obtain xy multiply each x value by its matched y value. The sums of all values (Σ) are given at the foot of the table.

Table 12.2 Dunlin weight and bill-length observations

Bill length (mm) x	Weight (g) y	x^2	y^2	xy
33.5	51	1122.25	2601	1708.5
38.0	59	1444	3481	2242
32.0	49	1024	2401	1568
37.5	54	1406.25	2916	2025
31.5	50	992.25	2500	1575
33.0	55	1089	3025	1815
31.0	48	961	2304	1488
36.5	53	1332.25	2809	1934.5
34.0	52	1156	2704	1768
35.0	57	1225	3249	1995
Σ 342	528	11752	27990	18119

Summarising the data:

$$n = 10$$

$\Sigma x = 342$ $\Sigma y = 528$

$(\Sigma x)^2 = 342^2 = 116964$ $(\Sigma y)^2 = 528^2 = 278784$

$\Sigma x^2 = 11752$ $\Sigma y^2 = 27990$

$$\Sigma xy = 18119$$

Readers will be aware, of course, that many of these terms can be displayed directly on a Scientific calculator after the input of the respective variable. If the calculator allows for the input of *two* variables, the correlation coefficient is also automatically calculated according to the formula below.

We now have all the terms needed to substitute in the formula for the calculation of the correlation coefficient:

$$r = \frac{n \cdot \Sigma xy - \Sigma x \cdot \Sigma y}{\sqrt{[n \cdot \Sigma x^2 - (\Sigma x)^2][n \cdot \Sigma y^2 - (\Sigma y)^2]}}$$

Substituting in the formula, we obtain:

$$r = \frac{(10 \times 18119) - (342 \times 528)}{\sqrt{[(10 \times 11752) - (116964)]\,[(10 \times 27990) - (278784)]}}$$

$$r = \frac{181190 - 180576}{\sqrt{[556] \times [1116]}} = \frac{614}{\sqrt{620496}} = \frac{614}{787.7} = 0.779$$

According to Table 12.1 there appears to be a strong and positive correlation between Dunlin weight and bill-length. We need to check that such a correlation is unlikely to have arisen by chance in a sample of 10 units. Appendix 3 gives the probability distribution of r. First we need to work out the degrees of freedom. In the case of a correlation coefficient they are the number of pairs of observations less 2, that is, $(n-2) = 8$ in our example. Consulting Appendix 3 we find that our calculated value of 0.779 exceeds the tabulated value at 8 df of 0.765 at $P = 0.01$. Our correlation is therefore statistically highly significant.

12.5 The coefficient of determination r^2

The square of the product moment correlation coefficient is itself a useful statistic and is called the **coefficient of determination**. It is a measure of the proportion of the variability in one variable that is accounted for by variability in another. In a perfect correlation where $r = +1$ or -1, a variation in one of the variables is exactly matched by a corresponding variation in the other. This situation is rare in biology because many factors govern relationships between variables in organisms. Thus, r^2 indicates to what extent other factors are independently affecting x and y. In Example 12.1 the coefficient of determination is $0.779^2 = 0.607$ or 60.7% It follows that about 40% of the weight of Dunlins is not accounted for by variation in bill-length. The weight variable is undoubtedly influenced by how much food a bird has in its crop or when it last defaecated, factors which are quite independent of bill length.

12.6 The Spearman Rank Correlation Coefficient r_s

When there is doubt that the rather rigorous conditions relating to the use of the product moment correlation coefficient are fulfilled, the use of the non-parametric alternative should be considered.

Example 12.2

In the British Trust for Ornithology's *Constant Effort Sites Scheme*, the number of juveniles captured at a site is a component of an index of productivity. Evidence that the number of juveniles captured does indeed reflect the number present would be very valuable. Such evidence would be forthcoming if the number of juveniles caught correlated well with the number of chicks ringed during exhaustive nest searches during the same season.

The numbers of blackbird nestlings ringed, and juveniles captured in the same season, over a 12 year study period at a woodland site are given in Table 12.3 (there are some extra columns in the table which are explained later). It is very unlikely that data of this kind are normally distributed, and so the Spearman rank correlation coefficient is appropriate to the problem.

Table 12.3 Numbers of Blackbird nestlings and juveniles ringed at a constant effort site over 12 successive years (*Ringing & Migration*, 1991, 12, 118)

Nestlings ringed	Rank	Juveniles captured	Rank	d	d²
16	12	16	12	0	0
9	8.5	8	9	−0.5	0.25
1	3	5	7	−4	16
0	1.5	3	3.5	−2	4
7	7	4	5	2	4
13	10	5	7	3	9
9	8.5	10	11	−2.5	6.25
15	11	9	10	1	1
3	4.5	3	3.5	1	1
5	6	2	2	4	16
3	4.5	5	7	−2.5	6.25
0	1.5	0	1	0.5	0.25
					$\Sigma\ d^2 = 64$

The basis of the test is to *rank* the values of each variable in ascending order. The smallest value of each is rank 1, and the largest is rank n, where n is the number of units (pairs of observations) in the sample. The actual values of the observations are then dispensed with and the *ranks* become the basic data used in the test. The ranks of each observation are included in Table 12.3. Note that in cases where the observations are equal in value, the ranks are averaged, as explained in Section 3.6. In the next column, headed *d*, is the arithmetic difference between the ranks of the two variables. If the figures in this column are summed they provide a "sum of differences". Since this is always zero, it is not a helpful statistic. However, it should be checked; if it does not equal zero, there is an error in the ranking. Of more use is the *sum of the squares of the differences*, Σd^2. This is obtained by squaring each of the 12 values of *d* and adding them up. Σd^2 is entered in the formula to calculate the correlation coefficient. The formula is:

$$r_s = 1 - \left[\frac{6\ \Sigma\ d^2}{n^3 - n} \right]$$

where n is the number of units in the sample and 6 is a constant peculiar to this formula.

Substituting,

$$r_s = 1 - \left[\frac{6 \times 64}{(12 \times 12 \times 12) - 12} \right] = 1 - \left[\frac{384}{1716} \right]$$

$$r_s = 1 - 0.224 = 0.776$$

There appears to be a strong correlation between the two variables. To test the significance of the relationship we consult Appendix 4. Entering Appendix 4 at $n = 12$ in the column headed

0.05 level of significance (two tailed test) we see that our calculated value of 0.776 exceeds the tabulated critical value of 0.591. We conclude there is a statistically significant correlation between the number of chicks ringed and the number of juveniles captured.

12.7 Restrictions and cautions

1. When observations of one or both variables are on an ordinal scale, or when it is suspected that data are not normally distributed, use the Spearman rank correlation coefficient. The number of units in a sample, that is, the number of paired observations should be between 7 and about 30. The ranking of over 30 observations is extremely tedious and is not commensurate with any marginal increase in accuracy. Where there are observations of equal value (tied observations) assign average ranks as described in Section 3.6.

2. When observations are measured on interval scales the use of the product moment correlation coefficient should be considered. Sample units must be obtained randomly and the data should be *bivariate normal*. This can be checked by inspection of a scattergram of the data. Do not attempt to measure correlations of *bimodal* data (see Section 5.5). If possible, separate the observations attributable to each mode, for example, male and female, before analysing correlations.

3. The relationship between two variables should be linear, not curved. A scattergram will show if this is the case. Certain mathematical transformations will 'straighten up' curved relationships to allow r to be calculated. The logarithmic transformation for counts and the arcsine transformation for proportions are commonly used.

4. Do not conclude that because two variables are strongly and significantly correlated that one is necessarily the *cause* of the other. It is always possible that some additional, unidentified, factor is the underlying source of variability in both variables.

5. Correlations measured in samples estimate correlations in the populations from which they are drawn. A correlation in a sample is not improved or strengthened by increasing the sample size; however, larger samples may be required to confirm the statistical significance of weak correlations.

CHAPTER 13
REGRESSION ANALYSIS

13.1 Introduction

In Section 4.7 we illustrate the relationship between weight and bill length of Dunlin by means of a scattergram. In presenting a scattergram it is often helpful to draw a line through the cloud of points in such a way that the average relationship is depicted. The line is called the **line of best fit**. It has been added to the scattergram in Fig. 13.1. A problem arises as to how to fit the best line through a cloud of points. If the scatter is not too great the line may be reasonably fitted 'by eye'. In most cases, however, it is necessary to replace such a subjective method by a more objective mathematical approach. The line so produced is called a **regression line**.

The regression line may be described in terms of a mathematical equation which defines the relationship between the x and y variables and we may use the equation to estimate or predict the value of one variable from a measurement of the other. We call this technique (i.e. fitting the best line to a scattergram from an equation relating x and y) **regression analysis**. Before we attempt to describe this important statistical technique we must first consider a little basic geometry.

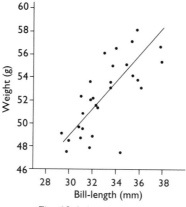

Fig. 13.1 A regression line

13.2 Gradients and triangles

The gradient of a hill slope is often expressed in such terms as 1 in 10. This means simply that for every 10 units of distance travelled in a horizontal plane an elevation of 1 unit in the vertical plane will result. The gradient, or slope, is symbolised by b and, in this case, is equal to $1 \div 10 = 0.1$. In the general case, the gradient is equal to an increment in y divided by an increment in x (Fig. 13.2).

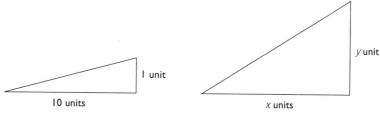

Fig. 13.2 Gradients of slopes: gradient = y units/x units

Knowing the value of b we may use it to calculate the height gained from a given distance moved horizontally, thus:

Vertical height gained = gradient x horizontal distance travelled

In the general case, this may be expressed as:

$$y = bx$$

If we wish to know the actual final height rather than just the height gained we must know the height of our starting point above some reference zero, say sea level. If this is symbolised by a units, the final height will be:

Final height = height of starting point + (gradient x horizontal distance travelled)

or, in the general case:

$$y = a + bx \text{ (see Fig. 13.3).}$$

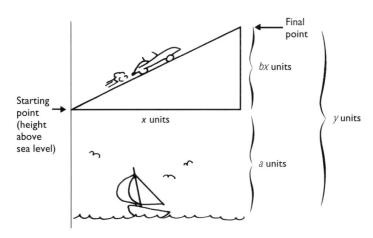

Fig. 13.3 Final height of car above sea level = a + bx where b is the gradient

The equation, $y = a + bx$, is known as the equation of a straight line or the **rectilinear equation**. Regression analysis is concerned with solving values of a and b in the equation from a set of bivariate sample data. We may then accurately fit a line to the scattergram and estimate the value of one variable from a measurement of the other. The quantities a and b are both **regression coefficients**. In common usage, however, the term 'regression coefficient' is taken to mean the slope of the regression line, b.

13.3 Dependent and independent variables

To this point we have not indicated which of two variables should be placed on the y (vertical) axis and which on the x (horizontal) axis. There is a convention which gives us a guideline in this respect. In many pairs of variables it is possible to discern that one of the variables is dependent on the other. For example, the number of warblers breeding in a county might depend on the length of hedgerow available. The notion that the converse might be true, that the length of hedgerow in some way depends on the number of warblers, is clearly ludicrous. Likewise, the height gained by the car in the last section depends on the horizontal distance moved. In these instances, the variable which describes the number of warblers and the height of the car is called the **dependent variable**. The other variable is called the **independent variable**. In a scattergram or during regression analysis, we conventionally place the dependent variable on the y-axis and the independent variable on the x-axis. We should note here that in identifying a dependent variable we are not necessarily admitting a *causal* relationship between the two. It is always possible that the two variables are independently related to a third, unidentified, variable.

Sometimes it is not possible to state which is the dependent and independent variable in a pair. Such is the case in the Dunlin weight / bill-length relationship of Example 12.1. If the regression line is to be used to estimate the value of one variable from a measurement of the other, then the variable which is used to estimate the other is placed on the x-axis; the variable *to be estimated* is placed on the y-axis. If, however, the line is calculated merely to describe the mathematical relationship between the two variables, and no dependent variable can be identified, then the choice of axis is arbitrary.

13.4 A perfect rectilinear relationship
By example, we may relate the rectilinear equation derived in Section 13.2 to the idea of dependent and independent variables

Example 13.1
An observer measures the length of a spring when different known weights are suspended from it. The following data are recorded:

Weight x (g)	Length of spring y (cm)
10	10
20	15
30	20
40	25
50	30

Clearly the length of the spring is dependent on the weight attached. We have no difficulty therefore in identifying weight as the x variable and length as the y variable. A scattergram drawn from these data is shown in Fig. 13.4a. We see that all points are in perfect alignment and that the regression line may be drawn through them without the need for a mathematical computation (Fig. 13.4b). When extrapolated downwards, the line cuts the y-axis at point a. This is the *intercept on the y-axis*. It represents the length of the spring in its unstretched state and is analogous to the *height above sea level* in the example in Section 13.2. We see from the scale on the y-axis that a has a value of 5 cm. Thus one of the quantities in the $y = a + bx$ equation has been determined.

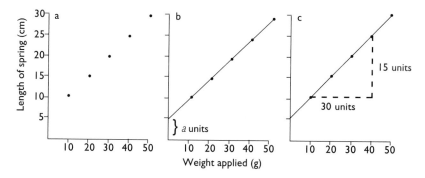

Fig. 13.4 A perfect linear relationship

To calculate the other, *b*, drop a vertical line from any point on the regression line and complete a right-angled triangle with a horizontal line as shown in Fig. 13.4c. The gradient, *b*, is the number of units on the *y* scale divided by the number of units on the *x* scale which correspond to the respective sides of the triangle. In Fig. 13.4c, $b = 15/30 = 0.5$.

We now have both quantities in the equation $y = a + bx$. Thus:

$$y = 5 + 0.5x$$

Using this equation we may now estimate the length of the spring when a weight of any size is attached. We can check this for a value we already know. An observation in the table above shows that a mass of 50 g produces a length of 30 cm. Does the equation bear this out?

$$y = 5 + (0.5 \times 50) = 5 + 25$$
$$y = 30 \text{ cm}$$

Clearly, we have obtained the correct result for the values of *a* and *b*. We may now confidently interpolate between the points and predict the length for, say, a weight of 25 g.

$$y = 5 + (0.5 \times 25) = 17.5 \text{ cm}$$

In this way we can build up a series of *y* values to produce a finely graded scale. If the scale were to be etched onto a stick aligned vertically beside the spring we would have a rudimentary spring balance.

Although it is safe to interpolate between points on the scattergram caution should be exercised in extrapolating far beyond the last point. If a 500 g weight is hung from the end of our spring it might well overload the capacity of the spring to return to its original length, and it is by no means certain that its stretched length is equal to that predicted by the equation.

As we have already noted, such perfect rectilinear relationships may be encountered in the physics laboratory but seldom in biology, where a cloud of points rather than a nice straight line is more usual.

13.5 The line of least squares

Fitting a regression line to a scattergram involves placing it through the points so that the sum of the vertical distances (**deviations**) of all points from the line is minimised. Because some deviations are negative and some positive, it is more convenient to utilise the sum of the *squares* of the deviations, Σd^2. In this way awkward negative signs are eliminated. The method of fitting the line is therefore known as the **method of least squares**. Figure 13.5 shows the line of least squares fitted to four points in which the squares of the vertical deviations are minimised.

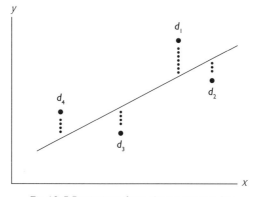

Fig 13.5 Deviations from the regression line

Any alternative position of the line (up or down, or with different slope) will increase Σd^2. Although it is possible to reduce the values of $d_1{}^2$ and $d_4{}^2$ by moving the line closer to these points, it is only at the expense of increasing $d_2{}^2 + d_3{}^2$. Since we are dealing with *squares*, the increase in $(d_2{}^2 + d_3{}^2)$ is not offset by the decrease in $(d_1{}^2 + d_4{}^2)$.

Fitting the line in this way, where Σd^2 for vertical deviations (deviations on the y-axis) is minimised, is called the **regression of y on x**. It allows for the estimation of y values from nominated values of x, and is usually referred to as **simple linear regression**. Regrettably, this title belies some rather strict conditions which apply to the application of least squares regression. The main conditions are:

(a) There is a linear relationship between a dependent y variable and an independent x variable which is implied to be *functional* or *causal*.

(b) The x variable is not a random variable but is under the control of the observer.

(c) The scatter of points is more or less evenly spread on each side of the regression line along its whole length (i.e. the scatter of points should not fan out towards one end of the line).

With point (b) in mind we can see that regression is a technique that is especially applicable to *experimental* situations in which, for example, the response of a y variable to *predetermined* quantities of time, temperature, dose of drug, application of fertiliser, and so on, is under investigation. Unfortunately, many of the situations in which ornithologists wish to fit a regression line to a scattergram do not meet these conditions. The better news is that, in some circumstances, the errors incurred by 'breaking the rules' are small enough to be ignored (Section 13.11); in other circumstances, there is an alternative method which can be utilised (Section 13.14).

To illustrate the application of simple linear regression, we choose a classical example which fulfils all the requirements of the method.

Example 13.2

An environmentally-friendly city council acquires a substantial area of derelict industrial land which it plans to restore as a recreational and educational area by encouraging birds and other wildlife. After decades of neglect, the land is nutritionally impoverished, and requires an application of fertilizer to establish regeneration of vegetation. The problem is to apply sufficient quantities of fertiliser to optimise vegetation growth and to avoid excess application which could lead to run-off and nutrient enrichment ("eutrophication") of nearby water courses.

Grass seed is sown uniformly over the area. Ten 1 m² plots are located randomly and a different weight of commercial fertiliser is applied evenly to each. Two months later the grass is carefully harvested from each plot, dried and weighed. The results of the experiment are tabulated below in Table 13.1.

Table 13.1 Weight of fertiliser applied

x variable: weight of fertiliser (g per sq. m)	25	50	75	100	125	150	175	200	225	250
y variable: yield of grass (g per sq. m)	84	80	90	154	148	169	206	244	212	248

A scattergram of these data (Fig. 13.7) reveals a roughly rectilinear relationship; within the limits of the experiment the yield of grass clearly depends on the amount of fertiliser applied and a functional relationship is presumed. Observations on the x-axis are under the control of the observer and so simple linear regression is appropriate.

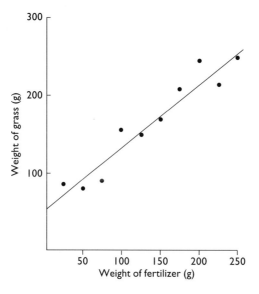

Fig 13.7 Regression of mass of grass yield on mass of fertilizer applied

13.6 Regression of weight of grass on weight of fertiliser applied

The information we require to calculate a and b is the same as that needed to calculate the correlation coefficient. In addition we must obtain the mean of the x observations ($\bar{x}$) and the mean of the y observations ($\bar{y}$). Using a calculator we determine:

$\bar{x}$	= 137.5	$\bar{y}$	= 163.5	n	= 10
Σx	= 1375	Σy	= 1635	Σxy	= 266650
$(\Sigma x)^2$	= 1890625	$(\Sigma y)^2$	= 2673225		
Σx^2	= 240625	Σy^2	= 304157		

The computation of a and b for the regression of y on x requires two steps. First the calculation of b, and then, using the value of b so derived, the calculation of a. The formula for the calculation of b is:

$$b = \frac{n\Sigma xy - \Sigma x \Sigma y}{n\Sigma x^2 - (\Sigma x)^2}$$

Substituting in the formula:

$$b = \frac{(10 \times 266650) - (1375 \times 1635)}{(10 \times 240625) - 1890625} = \frac{418375}{515625}$$

$$b = 0.8114$$

To calculate a we use the rectilinear equation, $y = a + bx$ or, rearranging, $a = y - bx$. We have derived a value for b, but which of the 10 possible values of x and y should we use to solve the equation for a? Since all points constitute a scatter it is most unlikely that any one single pair will be correct. The answer is to use the mean of x and the mean of y.

$$a = \bar{y} - b\bar{x}$$

We now have all the terms needed to solve for a:

$$a = 163.5 - (0.8114 \times 137.5)$$
$$a = 163.5 - 111.5675$$
$$a = 51.933$$

The regression equation may now be written out in full:

$$y = 51.933 + 0.8114x$$

We can use the equation to predict the yield of grass from any given application of fertiliser. Thus, for an application of 115 g we predict:

$$\text{Yield of grass (g)} = 51.93 + (0.8114 \times 115) = 145.2 \text{ g}$$

It is safe only to make predictions within the range of the initial x observations. The regression line undoubtedly curves off when the amount of fertiliser ceases to be the limiting growth factor. Indeed, the yield of grass may actually fall when the amount of fertiliser applied is so large as to become toxic; the surplus fertiliser is liable to wash out and pollute local water bodies.

The distribution of the observations on the x-axis is not normal because they span a range and are not clustered around an average value. We are therefore unable to test the *significance* of this relationship by means of the product moment correlation coefficient. The procedure for testing the significance of a regression line is described in Section 13.10.

13.7 Fitting the regression line to the scattergram

To fit a line to a scattergram we need to know the positions of two points on the line. The further apart the points are, the more accurately can the line be drawn. The points are calculated from the regression equation, $y = a + bx$. Two separate values of x are selected to derive corresponding y values from the equation. These become the coordinates of the two points.

Select a value of x that is close to the y-axis. In example 13.2 this could be 25 g. Calculate the corresponding y value from the equation:

$$y = 51.93 + (0.8114 \times 25) = 72.2$$

Mark a point at coordinates $x = 25$; $y = 72.2$. If the x-axis is scaled to start at zero, then the y-coordinate at $x = 0$ will of course equal a, the intercept. The point can be marked here on the y-axis. Select a second value of x, far from the y-axis, say 250 g. As before, calculate the corresponding y value from the equation:

$$y = 51.93 + (0.8114 \times 250) = 254.8$$

Mark a second point on the scattergram at coordinates $x = 250$; $y = 254.8$. Join the two points with a straight line but do not exceed the cloud of points by very much at each end.

13.8 The error of a regression line

From the regression equation of Example 13.2 we estimate that an application of 115 g of fertiliser results in a yield of 145.2 g of grass grass grown under standard conditions. Can we say how good an estimate this is? We recognise, first, that the regression line is derived from *sample* data and is therefore subject to the same sort of sampling error that occurs when we estimate a mean or any other population parameter. Suppose that we apply 115 g fertiliser to 1000 plots of 1 m²; that is to say, a *population* of plots. The yields of grass from the plots constitute a normal distribution of y values, all corresponding to the x value of 115 g. The mean of this distribution is the true population estimate and is the y coordinate through which the regression line passes without error. The standard deviation of the distribution is the **standard error (S.E.) of the estimate**. How the standard error is estimated is described below. Applying the same argument as in Section 9.4, we are 95% confident that the population value falls within $\pm(t \times S.E.)$ of an estimate based on a sample regression. Thus, in Example 13.2 the y-coordinate of the true regression line at $x = 115$ g is within $145.2 \pm (t \times S.E.)$. We obtain the value of t at the appropriate degrees of freedom from Appendix 1. Then a confidence interval can be placed vertically above and below the regression line at $x = 115$. By repeating this procedure for several values of x, a **confidence zone** is established on either side of the regression line. The upper and lower boundaries of the zone are not parallel to the regression line but fan out towards each end.

Before the standard error is obtained, a quantity called the **residual variance**, s_r^2 is first calculated using the formula:

$$s_r^2 = \frac{1}{n-2} \times \left(\text{sum of squares of } y - \frac{(\text{sum of products})^2}{\text{sum of squares of } x} \right)$$

We solve, in turn, each item within the large brackets.

(a) The sum of products (SP) is a quantity involving both x and y and is calculated from the formula:

$$SP_{x,y} = \Sigma xy - \frac{\Sigma x \Sigma y}{n}$$

Inserting the terms we derived in Example 13.2

$$SP_{x,y} = 266650 - \frac{1375 \times 1635}{10} = 41837.5$$

We need the square of this quantity, thus:

$$(SP_{x,y})^2 = 41837.5^2 = 1750376406$$

(b) The sum of squares (SS) of y is given by (Section 6.6):

$$SS_y = \Sigma y^2 - \frac{(\Sigma y)^2}{n} = 304157 - \frac{2673225}{10} = 36834.5$$

(c) In the same way, the sum of squares of x is given by (Section 6.6):

$$SS_x = \Sigma x^2 - \frac{(\Sigma x)^2}{n} = 240625 - \frac{1890625}{10} = 51562.5$$

(d) Substitute these numbers in the formula above for residual variance:

$$s_r^2 = \frac{1}{8} \times \left(36834.5 - \frac{1750376406}{51562.5} \right) = \frac{1}{8} \times 2887.806 = 360.98$$

To calculate the standard error of a point on the regression line corresponding to an estimate of y (designated y') from a stated value of x (designated x'), the formula is:

$$S.E. = \pm \sqrt{s_r^2 \times \left[\frac{1}{n} + \frac{(x' - \bar{x})^2}{SS_x}\right]}$$

Using our regression equation from example 13.2, a value of x' of 115 g of fertiliser gives an estimate, y', of 145.2 g yield of grass. Substituting these values in the equation for the standard error (with SS_x derived in (c) above):

$$S.E. = \pm \sqrt{360.98 \times \left[\frac{1}{10} + \frac{(115 - 137.5)^2}{51562.5}\right]}$$

$$= \pm \sqrt{360.98 \times 0.10982}$$

$$= \pm 6.296$$

To convert the standard error to a 95% confidence interval we multiply by the appropriate value of t (see Section 9.4). Consulting Appendix 1 at $P = 0.05$ the tabulated value of t at $(n - 2) = 8$ df is 2.306. The 95% confidence limits (C.L.) about y' are therefore $y' \pm t \times S.E.$

$$95\% \text{ C.L.} = 145.2 \pm 2.306 \times 6.296$$

$$= 145.2 \pm 14.519$$

$$= 159.72 \text{ (upper limit) and } 130.681 \text{ (lower limit)}$$

We are therefore 95% confident that if we were to replicate the experiment a very large number of times (thereby producing a *population* regression line), the point on the line corresponding to $x = 115$ will fall between $y = 159.72$ and $y = 130.681$.

Computation of the 95% confidence limits for several values of x over the range of the x-scale will enable the 95% confidence zone to be plotted. Because of the fanning out effect, it is not possible to define the zone from only two values of x, as it was when drawing the regression line itself (Section 13.7). The 95% confidence zone of our regression line is shown as the *inner* zone in Fig. 13.8.

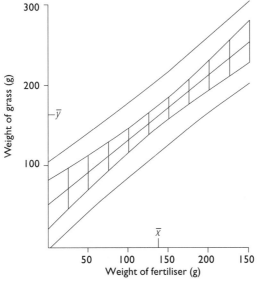

Fig. 13.8 Confidence zones of a regression line.

13.9 Confidence limits of an individual estimate

The 95% confidence zone described in the previous section is the confidence zone for the line as a whole. Predicted values of y for *individual* values of x are subject to an additional source of error, namely scatter about the regression line. We must therefore establish a second confidence zone whose limits are further out from the limits of the confidence zone of the regression line.

If the 95% confidence limits for the regression line have been calculated, those for individual estimates are determined by adding 1 in the square root term in the equation used for calculating the limits of the regression line. The zone may be determined by similarly working out the confidence limits for several values of x over the range of the x-scale. Thus:

$$95\% \text{ C.L.} = y' \pm t \times \sqrt{s_r^2 \times \left[1 + \frac{1}{n} + \frac{(x' - \bar{x})^2}{SS_x} \right]}$$

Using the same values as in the previous section, the 95% confidence limits about the predicted yield from a single application of 115 g fertiliser are:

$$95\% \text{ C.L.} = 145.2 \pm t \times \sqrt{360.98 \times (1 + 0.10982)}$$

$$= 145.2 \pm (2.306 \times 20.016)$$

$$= 145.2 \pm 46.156$$

$$= 191.356 \text{ g (upper limit) and } 99.04 \text{ g (lower limit)}$$

What this means is that if we were to make a *single* application of 115 g of fertiliser to a 1 m² plot under standard conditions we are 95% confident that the yield of grass will be between 191.36 g and 99.04 g. The limits are shown as the *outer* zone in Fig. 13.8 and represent a considerable extension to the breadth of the confidence zone of the regression line.

13.10 The significance of the regression line

If the scatter of points about the regression line is considerable, and the value of the regression coefficient, b, is low, it may be necessary to test the significance of the regression. This will tell us the probability that there is a true linear relationship between the x and y variables in the parent population.

If there is no relationship, the value of b is zero. That is, the slope of the regression line is parallel to the x-axis. The test for significance is therefore a test for a significant departure of the value of b from zero. The test depends on the computation of a t value (at $n - 2$ df) from the values of b and the residual variance, s_r^2, whose derivation was described in Section 13.8. The steps involved in the calculation of t are as follows. The three quantities needed are s_r^2 (360.98 from Section 13.8); the sum of squares of x (51562.5 from Section 13.8); and the value of the slope, b (0.8114 from Section 13.6).

Step 1: Work out the standard error (S.E.) of b from the formula:

$$\text{S.E.}_b = \sqrt{\frac{\text{residual variance}}{\text{sum of squares of } x}}$$

$$= \sqrt{\frac{360.98}{51562.5}} = 0.08367$$

Step 2: Estimate t from:

$$t = \frac{b}{\text{S.E.}_b} = \frac{0.8114}{0.08367} = 9.698 \text{ with } (n-2) = 8 \text{ df.}$$

Step 3: Compare the calculated value with the tabulated value at 8 df in Appendix 1. The calculated value greatly exceeds the value of 3.355 at $P = 0.01$. The linear relationship between the y and x variables is highly significant.

Alternatively, significance may be tested by analysis of variance (Section 15.10).

13.11 When there is no independent variable

In Section 13.5 we said that one of the conditions applying to the use of simple linear regression was that the x variable is not a random variable, but is under the control of the observer.

There are circumstances when ornithologists may wish to fit a line to a scattergram and make predictions when this condition is not met. Fortunately, statisticians allow us to use simple linear regression provided always that the variable *from which estimations are made* is assigned to the x-axis.

Example 13.3

In the studies of the feeding ecology of seabirds, it is desirable to know the sizes of fish prey taken. Fish contain in their heads a pair of "stones" called *otoliths* ("ear stones") which assist in orientation. They resist digestion in a bird's crop and often appear in regurgitated material after the fish and its bones have dissolved. Because each species of fish has its own characteristically shaped otoliths, the identity of the fish prey can be deduced. If it can also be demonstrated that there is a positive relationship between otolith length and fish length, the size of the fish prey taken can be estimated from the size of the otoliths recovered in regurgitated material.

A sample of fresh fish of known species of varying length are measured, and then the otoliths from each fish are removed and also measured. This generates a set of bivariate data which may be plotted as a scattergram. However, there is no clear dependent variable. Both fish length and otolith length are dependent variables, related no doubt to some other underlying genetically controlled variable. However, because we eventually wish to make estimations of fish length from lengths of otoliths recovered in regurgitates, it is *otolith length which is assigned to the x variable*. By way of example, the equation that has been worked out by simple linear regression for sandeels (*Seabird*, 1987, 10, 71) is:

$$\text{Sandeel length} = 27.92 + (44.01 \times \text{otolith length})$$
$$y = a + b \qquad x$$

Thus, an otolith of length 1.5 mm, and recognised by its shape as belonging to a sandeel, is found in a Fulmar regurgitate. The estimated length of the sandeel from which the otolith came is:

$$\text{Sandeel length} = 27.92 + (44.01 \times 1.5) = 93.9 \text{ mm.}$$

Confidence limits to the regression line, and to individual estimates, may be worked out exactly as described in Sections 13.8 and 13.9.

Sometimes ornithologists wish to fit a line to a scattergram comprising two dependent variables to show the "average" relationship without necessarily wishing to make predictions of one value from another. In this case, proceed with the alternative regression line described in Section 13.14.

13.12 Dealing with curved relationships

We noted in Section 13.5 that one of the conditions applicable to simple linear regression is that there is a *linear* relationship between the x and y variables. Many relationships in biology are not linear but exhibit curved lines of best fit. The usefulness of regression is greatly increased when curved relationships are 'straightened up' by *transformation*. It then becomes possible to undertake regression analysis and to estimate a value of one variable from another.

Example 13.4

In a study of Great Skua productivity, a population of chicks is monitored from hatching and wing lengths at known ages recorded. The results are shown in the scattergram in Fig. 13.9(A). In this case the curve is straightened by transforming the wing length measurements to their logarithms, i.e. replacing y by $\ln(y)$. Figure 13.9(B) shows the relationship as approximately rectilinear. (In this instance we have used natural logarithms, but $\log_{10}$ will also exert the straightening effect).

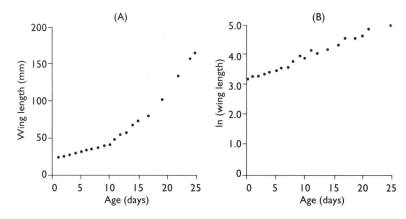

Fig. 13.9 A logarithmic transformation: A, untransformed; B, transformed (courtesy R. W. Furness).

The transformation appears to be satisfactory. We may therefore proceed with regression analysis provided that all values of y are replaced by their logarithms. By way of illustration we take 10 of the Great Skua wing measurements (Table 13.2).

Table 13.2. Great Skua wing length measurements

Age (days) x	Wing length (mm) y	ln(wing length) y'
2	26.0	3.26
4	29.2	3.37
6	33.9	3.52
8	37.0	3.61
10	41.6	3.73
13	57.4	4.05
16	73.7	4.30
19	81.5	4.40
22	105.0	4.65
25	164.0	5.10

Summarising the data, where y' is ln(wing length):

$$n = 10$$

Σx	=	125		$\Sigma y'$	=	39.99
$(\Sigma x)^2$	=	15625		$(\Sigma y')^2$	=	1599.2
Σx^2	=	2115		$\Sigma y'^2$	=	163.2
$\bar{x}$	=	12.5		$\bar{y}'$	=	4.00

$$\Sigma xy = 542.15$$

Substituting in the equation to solve b (Section 13.6), we derive:

$$b = \frac{n.\Sigma xy' - \Sigma x.\Sigma y'}{n.\Sigma x^2 - (\Sigma x)^2}$$

$$b = \frac{(10 \times 542.15) - (125 \times 39.99)}{(10 \times 2115) - (15625)}$$

$$b = 0.0765$$

As before, we may substitute for a:

$$a = \bar{y}' - b\bar{x}$$
$$a = 4.00 - (0.0765 \times 12.5)$$
$$a = 3.04$$

We now have the values for the quantities a and b. The relationship between wing length and age thus becomes:

$$\text{ln(wing length)} = 3.04 + 0.0765 \times \text{age}$$

To estimate the length of a wing at age 15 days:

$$y' = 3.04 + (0.0765 \times 15) = 4.19.$$

The wing length is a number, in mm, whose natural logarithm is 4.19 (i.e. antilog 4.19), namely 66.0 mm.

In practice, it would be more usual to want to estimate a chick's age from its wing length. Provided that no more than an approximate estimate of age is required, it is possible to recast the data, assigning the age variable to the y axis and the wing length variable to the x axis and to recalculate the regression of y on x. Note, however, that the rule of assigning the dependent variable to the y axis is broken, and that is why estimates will only be approximate. They will nevertheless be adequate for practical purposes. The point to note is that, without the transformation of the wing length variable, it would not have been possible to undertake regression at all.

13.13 Transformation of both axes

Logarithmic transformation of one or both of the axes of a curvilinear relationship will usually result in an adequate straightening. But how do we decide whether one of the two axes, or both, should be transformed? The answer may be: by trial and error. If a scattergram of untransformed bivariate data is curved, prepare scattergrams with first one axis transformed, then the other, then both, until the best result is obtained. The best outcome can usually be assessed by visual inspection of the scattergrams. If the scatter is considerable and there is doubt, then a more objective method is to derive the **coefficient of determination** r^2 (see Section 12.5) for each alternative transformation. The one which gives the largest value of r^2 is the one to use for regression.

Example 13.5

Island biogeographic theory tells us that the number of species on islands increases with the size of the island, but not in rectilinear fashion. Table 13.3 shows the number of breeding bird species (S) on 11 small islands of given area (A) in Shetland. Included in the table are the natural logarithms of the number of species, $\ln(S)$, and of the areas $\ln(A)$. Because the number of species is clearly dependent on the area of the island, we assign S to the y axis and A to the x axis.

Table 13.3 Number of bird species on islands of different size

Island	(A) Area x (ha)	ln area x'	(S) No. species y	ln(No. species) y'
Hascosay	270	5.60	24	3.18
Bigga	76	4.33	22	3.09
Samphrey	72	4.28	21	3.04
Linga	40	3.69	13	2.56
Brother Isle	34	3.53	15	2.71
Uynarey	21	3.04	16	2.77
Orfasay	10	2.30	9	2.20
Wether Holm	4	1.39	6	1.79
Kay Holm	2	0.69	7	1.95
Little Holm	0.75	−0.29	3	1.10
Sinna Skerry	0.25	−1.39	3	1.10

Fig. 13.10 A–D shows, respectively, scattergrams of:

A: number of species *v.* area of island

B: ln (number of species) *v.* area of island

C: number of species *v.* ln(area of island)

D: ln(number of species) *v.* ln(area of island).

In each case the line shown has been fitted "by eye".

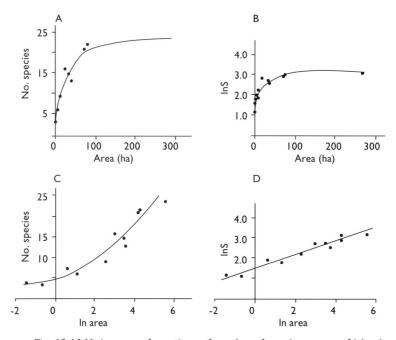

Fig. 13.10 Various transformations of number of species v. *area of island.*

It is clear from Fig. 13.10 D that the "double log." plot produces a satisfactory rectilinear transformation. It is therefore upon the logarithms of both S and A that the regression analysis should be undertaken. A regression of y' (ln$S$), on x' (ln$A$), enables a prediction of the number of species that would be expected on an island of given area. Because many ecologists regard nature reserves as "islands" in the biogeographical sense, predictions of this nature are helpful in the ranking of sites for conservation purposes. "Islands" which are found by field observations to have more than the predicted number of species may be regarded as more important.

Proceeding with linear regression on the transformed observations, the essential data are summarised (remember we are designating ln$x$ as x' and ln$y$ as y'):

$$n = 11$$

$$\Sigma x' = 27.17 \qquad\qquad\qquad \Sigma y' = 25.5$$
$$(\Sigma x')^2 = 738 \qquad\qquad\qquad \Sigma x'^2 = 113.5$$
$$\bar{x}' = 2.47 \qquad\qquad\qquad \bar{y}' = 2.32$$
$$\Sigma x'y' = 78.68$$

Solving for b (Section 13.5):

$$b = \frac{(11 \times 78.68) - (27.17 \times 25.5)}{(11 \times 113.5) - (738)} = 0.338$$

Solving for a':

$$a' = (\bar{y}' - b\bar{x}') = 2.32 - (0.338 \times 2.47) = 1.49$$

Therefore

$$y' = 1.49 + 0.338x'$$

Thus,

$$\text{ln(number of species)} = 1.49 + [0.338 \times \text{ln(area of island)}]$$

To estimate the number of species S on an island of area 120 ha:

$$\text{ln}(S) = 1.49 + (0.338 \times \text{ln}120) = 3.11$$

The estimated number of species is a number whose natural logarithm is 3.11 (i.e. antilog 3.11), namely, 22.4. Since species can only occur in whole numbers, this is rounded to 22.

An equation in the form ln$y' = $ ln$a' + b$.lnx' may be back-transformed by taking antilogs on each side of the equation to derive:

$$y = a \times x^b, \text{ where } a \text{ is antilog } a' = 4.44$$

Thus, the number of species expected on an island of 120 ha is:

$$S = 4.44 \times 120^{0.338}$$
$$= 4.44 \times 5.04 = 22.4$$

13.14 An alternative line of best fit

In section 13.5 we outline some of the conditions that have to be met if regression by least squares is to be correctly applied. One of the important conditions is that the x variable is not a *random variable* but is fixed or controlled by the observer. In many instances where ornithologists wish to place a line of best fit through points on a scattergram, this condition is not met. In

Example 12.1 we analyse bivariate data which reveal a strong correlation between the bill length and weight of Dunlins. Here, both variables – bill length and weight – are random, dependent variables. Neither variable is under the control of the observer. If we wish to make predictions about the weight of a Dunlin from the length of its bill, we should proceed as described in Section 13.11. However, if we merely wish to place a line through the scattergram to depict the "average" relationship, estimates of a and b based on least squares regression are likely to be statistically biased. In circumstances like this, an alternative form of regression known as **Model 2** is applied. There are a number of versions of Model 2 regression (see Sokal and Rohlf, 1981, Section 14:13). By far the simplest to apply is that known as **reduced major axis regression**.

The slope of the regression line (symbolised b' to distinguish it from least squares regression is given by:

$$b' = \pm \frac{\text{standard deviation of } y \text{ observations}}{\text{standard deviation of } x \text{ observations}}$$

The notation '$\pm$' is placed in front of the ratio because the formula does not give the *sign* of the slope. Standard deviations, it should be noted, are always positive. The sign of the slope, + or –, is decided by inspection of the scattergram. The intercept, a', is estimated in the usual way from $a' = (\bar{y} - b'\bar{x})$.

Two further points should be held in mind when applying Model 2.

(i) There is no requirement that sampling units should be obtained randomly. Indeed, it is preferable to select items which span the available range for measurement.

(ii) Because both variables are random variables, there is no clear dependent variable. The x and y axes are therefore assigned arbitrarily.

Example 13.6
The bill length and weight data of the Dunlins used in Example 12.1 are reproduced in Table 13.4.

Table 13.4 Dunlin weight and bill-length observations

Bill length (mm) x	Weight (g) y
33.5	51
38.0	59
32.0	49
37.5	54
31.5	50
33.0	55
31.0	48
36.5	53
34.0	52
35.0	57

Using a scientific calculator we determine:

Mean of x data, $\bar{x} = 34.2$; standard deviation of x data $s_x = 2.486$
Mean of y data, $\bar{y} = 52.8$; standard deviation of y data $s_y = 3.521$

$$b' = \pm \frac{s_y}{s_x} = \frac{3.521}{2.486} = 1.416$$

From an inspection of the data it is obvious that as the length of the bill increases, so too does the weight of the bird. The sign of the slope is therefore positive. Solving for a':

$$a' = (\bar{y} - b\bar{x}) = 52.8 - (1.416 \times 34.2) = 4.373$$

The equation is:

$$y = 4.373 + 1.416x$$

This equation defines the mutual slope of the two random variables (bill length and weight). It should be used to draw the line of 'best fit' through a scattergram of such data.

The *significance* of the line may *not* be tested by the method described in Section 13.10. A 'not significant' result would suggest that b' does not depart from zero, implying that one of the standard deviations is zero. That is not possible. Significance is tested by analysis of variance (Section 15.10).

13.15 Restrictions and cautions

1. A regression equation can be derived for any set of bivariate data simply by substituting them in the formulae. Because we can undertake a particular mathematical treatment, however, it does not mean that we *should*. The use of regression analysis should be restricted to cases where it is necessary to place a *best line* through a cloud of points and, in particular, where the estimation of one variable from a measurement of the other is required. If all that is required is a measure of the strength of a relationship between two variables the use of the correlation coefficient may be more appropriate.

2. Before proceeding with the mathematics of regression, always draw a scattergram of the data to see what it *looks* like. From this you can decide if two important conditions are met: (i) Are the points roughly linear, not curved? and (ii) is the scatter of points around the line reasonably even over the whole length of the line, not fanning out towards one end?

3. If the scattergram suggests a *curved* relationship then transformation of one or other or both axes will usually straighten up the line. We have given examples of logarithmic transformations, but arcsine transformations for proportions or square-root transformations may prove effective.

4. The correlation coefficient r has no meaning in regression unless sample units have been obtained randomly and observations are normally distributed on both axes. If this is the case the cloud of points in a scattergram assumes an elliptical shape – see Section 12.4. However, the *coefficient of determination* r^2 is a useful number. It tells us the proportion of the variability in the y observations that can be accounted for by variability in the x observations. It is *not* an index of significance.

5. The *significance* of a regression line is determined by testing for a significant departure of b, the slope, from zero. Alternatively, it may be determined by analysis of variance (Section 15.10). Analysis of variance is the only means of testing significance in *Model 2* regression.

6. In reporting the results of your regression analysis remember to say which method you use.

CHAPTER 14
COMPARING AVERAGES

14.1 Introduction

Ornithologists often wish to compare the average value of some variable in two samples. The problem typically resolves into two steps:

(i) is the observed difference between the average of two samples *significant*, or is it due to chance sampling error that we describe in Section 10.3?

(ii) if a difference between the average of two samples is indeed significant, what is the extent of the difference?

We have already dealt with the second step in Sections 9.6 and 9.7. In this chapter we outline methods which test the significance of a difference between the average of two samples. In each method we describe, the null hypothesis is similar:

H_0: samples are drawn from populations with identical averages and any observed difference between the samples is due to sampling error.

The alternative hypotheses are:

H_1 (two-tailed test): samples are drawn from populations with different averages; an observed difference between samples cannot be accounted for by sampling error.

Or,

H_1 (one-tailed test): a nominated sample is drawn from a population with a larger average than that from which the other is drawn.

Tests for difference between sample averages may be non-parametric or parametric. Non-parametric tests convert observations to ranks and compare sample distributions, essentially their *medians*. Parametric tests use actual observations and compare *means*. We describe non-parametric methods first and then proceed to parametric methods.

14.2 Matched and unmatched observations

When analysing bivariate data such as correlations, a single sample unit gives a pair of observations representing two different variables. The observations comprising a pair are uniquely linked and are said to be *matched*. It is also possible to obtain a pair of observations of a *single* variable which are matched. For example, the weights of 10 birds at one site and the weights of another 10 at another site are unmatched. However, the weights of 10 birds on one day and the weights of the same 10 birds, identified by their ring numbers a week later, are matched because the weight from one day can be paired off with that from the other. It is possible to conduct a more sensitive analysis if the observations are matched.

14.3 The Mann-Whitney *U*-test for unmatched samples

The Mann-Whitney *U*-test is a non-parametric technique for comparing the medians of two unmatched samples. It may be used with as few as four observations in each sample. Because the values of observations are converted to *ranks*, the test may be applied to variables measured on ordinal or interval scales. Moreover, because the test is distribution-free, it is suitable for data which are not normally distributed, for example, counts of things, proportions or indices. Sample sizes may be unequal. The task is to compute a test statistic U which is compared with tabulated critical values (Appendix 5).

Example 14.1

The wing lengths (mm) of 6 male and 8 female Great Tits are presented below. They are listed in increasing order for convenience.

| Sample 1 Males: | 73 | 74.3 | 75 | 75.3 | 75.5 | 75.8 | | |
| Sample 2 Females: | 71 | 71.5 | 72 | 72.4 | 73.5 | 74 | 74.3 | 75.2 |

A first glance suggests that males are longer-winged than females, but there is some overlap. It is reasonable to ask if the apparent difference is statistically significant or if it arises from sampling error. (Note: the fact that the data suggest that males are larger does *not* mean that our test is *one-tailed*. Before seeing the data we had no preconception of which might be the larger; we merely wish to know if there is a difference). The procedure for using the test is as follows.

1. List all observations in both samples in ascending order, assigning them ranks. Where there are tied ranks, the average is assigned, as explained in Section 3.6. Distinguish between samples by underlining one of them; we have underlined observations in Sample 1.

| Observation: | 71 | 71.5 | 72 | 72.4 | 73 | 73.5 | 74 | 74.3 | 74.3 | 75 | 75.2 | 75.3 |
| Rank: | 1 | 2 | 3 | 4 | 5 | 6 | 7 | 8.5 | 8.5 | 10 | 11 | 12 |

| Observation: | 75.5 | 75.8 |
| Rank: | 13 | 14 |

2. Sum the ranks for each sample; that is, sum the ranks of the underlined and non-underlined separately. Let R_1 = sum of ranks of Sample 1 and R_2 = the sum of ranks of Sample 2.

$$R_1 = 5 + 8.5 + 10 + 12 + 13 + 14 = 62.5$$
$$R_2 = 1 + 2 + 3 + 4 + 6 + 7 + 8.5 + 11 = 42.5$$

3. Calculate the test statistics U_1 and U_2 from:

$$U_1 = n_1 n_2 + \frac{n_2(n_2 + 1)}{2} - R_2$$

$$U_1 = 6 \times 8 + \frac{8 \times 9}{2} - 42.5$$

$$= 48 + 36 - 42.5 = 41.5$$

$$U_2 = n_1 n_2 + \frac{n_1(n_1 + 1)}{2} - R_1$$

$$U_2 = 6 \times 8 + \frac{6 \times 7}{2} - 62.5$$

$$= 48 + 21 - 62.5 = 6.5$$

4. Check at this point that $U_1 + U_2 = n_1.n_2$

$$41.5 + 6.5 = 6 \times 8$$

This being the case, proceed to Step 5.

5. Select the smaller of the two U values (i.e. $U_2 = 6.5$ in this example) and compare it with the value in the table (Appendix 5) for the appropriate values of n_1 and n_2 (6 and 8 in this example). In the table, the critical value at $n_1 = 6$ and $n_2 = 8$ is 8. Our smaller value of U is *less* than the critical value. The null hypothesis is therefore rejected. There is a statistically significant difference between the medians ($U = 6.5$; $P < 0.05$, Mann-Whitney U-test).

14.4 Restrictions and cautions in using the Mann-Whitney U-test

1. The Mann-Whitney U-test may be applied to interval data (measurements), to counts of things, derived variables (proportions and indices) and to ordinal data (abundance scales etc.). It may be used with as few sampling units in each sample as 2 and 8, 5 and 3, or 4 and 4. However, with samples as small as these, there must be no overlap of observations between the two samples for H_o to be rejected. Since this can be determined by inspection of the data it is hardly worth proceeding with the test.

2. Note that, unlike some test statistics, the calculated value of U has to be *smaller* than the tabulated critical value in order to reject H_o.

3. The test is a test for difference in *medians*. It is a common error to record a statement like "the Mann-Whitney U-test showed there is a significant difference in means". There is, however, no need to calculate the medians of each sample to do the test.

4. Although there is no requirement for the observations in the samples to be normally distributed, the test does assume that the two distributions are similar. It is therefore not permissible to compare the median of a positively skewed distribution with that of a negatively skewed one. Since it is usually impossible to identify a frequency distribution in small samples the point is largely academic. Nevertheless, if it is known from other studies that the two samples have been drawn from populations which have fundamentally different distributions then the test should not be used.

14.5 More than two samples – Kruskal-Wallis test

Sometimes ornithologists need a test which compares the averages of several samples. Suppose that an ornithologist wishes to investigate the dispersion of shearwater burrows in a colony. The number of burrows counted in five randomly placed quadrats in each of four sectors (north, south, east and west) of a large colony is recorded. This generates four samples (one from each sector) of five observations in each. It is possible to compare the averages of the four samples using the Mann-Whitney U-test, but the test would have to be repeated six times to compare N with E, N with W, N with S and so on for all combinations. Apart from being extremely tedious, there is an important statistical reason for avoiding *multiple comparisons* of this type. We explain the reason fully in Section 15.1. The Kruskal-Wallis test is a simple non-parametric test for comparing the medians of three or more samples. Observations may be interval scale measurements, counts of things, derived variables or ordinal ranks. Samples do not have to be of equal size. The method of performing the test is explained in the following example.

Example 14.2

An ornithologist counts the number of shearwater burrows in five random quadrats in four sectors (N, S, E, W) of a large colony. Are there differences between the average count in each sector? The steps in performing the Kruskal-Wallis test are as follows and relate to table 14.1.

Table 14.1 Number of shearwater burrows (rank score in brackets)

	N	S	E	W	
	27 (12)	48 (16)	11 (6)	44 (15)	
	14 (7)	18 (9.5)	0 (1)	72 (19)	
	8 (4.5)	32 (13)	3 (2)	81 (20)	
	18 (9.5)	51 (17)	15 (8)	55 (18)	
	7 (3)	22 (11)	8 (4.5)	39 (14)	
n	5	5	5	5	$N = 20$
R	36	66.5	21.5	86	
R^2	1296	4422.25	462.25	7396	
R^2/n	259.2	884.45	92.45	1479.2	$\Sigma(R^2/n) = 2715.3$

1. Tabulate the observations in columns for each sample N, S, E, and W. Assign to each observation its *rank within the table as a whole*. If there are tied ranks assign to each its average rank as described in Section 3.6. Place each rank in brackets beside its observation.

2. Write the number of observations n in each sample (five in each in this example) in a line underneath its respective column. Add these up to obtain N, the total number of observations (20 in this case).

3. Sum the ranks of the observations in each sample and write them (R) in the next line under n.

4. Square the sum of ranks and write these (R^2) in a line under R.

5. Divide each value of R^2 by its respective value of n. Write this (R^2/n) in the bottom line under R^2. Add up the separate values of R^2/n to obtain $\Sigma R^2/n$.

The results of steps 1 to 5 are shown in Table 14.1.

6. The test statistic, K, is obtained by multiplying $\Sigma(R^2/n)$ by a

factor $\dfrac{12}{N(N+1)}$ and then subtracting $3(N+1)$.

$$K = \left[\Sigma(R^2/n) \times \frac{12}{N(N+1)} \right] - 3(N+1)$$

$$K = \left[2715.3 \times \frac{12}{20(21)} \right] - 3(21)$$

$$K = 14.58$$

7. Compare K with the tabulated distribution of χ^2 (Appendix 2). Work out the degrees of freedom from the number of samples less one: $(4 - 1 = 3)$ in this example. At 3df our calculated value

of 14.58 exceeds the tabulated value of 11.34 at $P = 0.01$. We reject the null hypothesis and conclude that there is a highly significant difference between the average number of burrows in the sectors.

It should be remembered that the test is applied to the samples *as a group* and that we are confident only that there are differences within the group as a whole. We should therefore be cautious in making inferences about differences between particular pairs of samples, or between one sample and the others. However, it is safe to assume that *at least* there is a significant difference between the two samples which have the highest and lowest sum of ranks. Inspecting the table of data we note that these are sample W and sample E, respectively. We infer that these two, at least, are significantly different from each other.

14.6 Restrictions and cautions in using the Kruskal-Wallis test

1. Apply the test to compare the locations (averages) of three or more samples. If there are only three samples there should be more than five observations in each sample.

2. The test statistic, K, is compared with the distribution of χ^2. This does not mean however that observations have to be frequencies. Data may be on any scale of measurement that allows ranking and need not be normally distributed.

3. If the outcome of the test suggests rejection of H_0, be cautious about making *a posteriori*, that is, unplanned comparisons between samples, other than between the two with the highest and lowest sum of ranks. Having said that, 'common sense' should be used. It may be perfectly obvious from an inspection of the distribution of ranks that, for example, a particular sample stands out from the remainder.

14.7 The Wilcoxon's test for matched pairs

The Wilcoxon's test for matched pairs is a simple non-parametric test for comparing the medians of two matched samples. It calls for the calculation of a test statistic T whose probability distribution is known. In that test, one observation in a matched pair is subtracted from the other. Observations must therefore be measured on an *interval* scale. It is not possible to use the test for ordinal measurements such as abundance scales.

Example 14.3

An ornithologist working at a south coast reed swamp wishes to know if the habitat is used by migrating Reed Warblers for 'fattening up' before taking off on migration. Birds arrive in numbers during August and stay until at least the end of September. Birds weighed in September seem to be heavier than those in August. To test the significance of the relationship one could compare the median of the sample weighed in August with that of a sample in September using the Mann-Whitney U-test. If it were possible however to recapture some of the birds weighed in August during September, the two weights of each birds would constitute a *matched pair*. As we noted in Section 14.2, it is then possible to conduct a more sensitive analysis with matched data than with unmatched data. The Wilcoxon's test is appropriate for this problem.

The recorded weights are shown in Table 14.2. The null hypothesis H_0 is that there is no difference in median weights between the two sets of data. The alternative hypothesis, H_1 is that there is a difference, but with no prediction as to which way that difference will lie (i.e. a two-tailed test).

Table 14.2 Matched weights in two bird samples

Weight of bird weighed in August, Sample A (g)	Weight of same bird weighed in September, Sample B (g)
10.3	12.2
11.4	12.1
10.9	13.1
12.0	11.9
10.0	12.0
11.9	12.9
12.2	11.4
12.3	12.1
11.7	13.5
12.0	12.3

If H_0 is true, we have two expectations, namely:

(i) the number of weight *gains* will be matched by a similar number of weight *losses*. If H_1 is true, there will be more of one than the other.

(ii) the *sizes* of any weight changes will be balanced evenly between gains and losses; if H_1 is true, there will be a tendency for the larger changes to be in one direction.

The Wilcoxon's test for matched pairs quantifies both the direction and the magnitude of all the changes in a set of matched pairs. The steps in carrying out the test are as follows:

1. For each matched pair, subtract the value of observation A from the value of observation B. The answer is the difference, d. The value of d will have a negative sign if A is larger than B.

2. Rank the values of d according to their absolute values. That is to say, ignore any plus or minus signs for the moment. Ignore any instances in which $d = 0$. If any ranks are tied, assign the average of the ranks exactly as described in Section 3.6.

3. Assign to each rank a '+' or a '−' sign corresponding to the sign of d.

4. Sum the ranks of the "plus" values and the "minus" values separately. The results of steps 1 to 4 are given in Table 14.3.

    Sum of "minus" ranks: $1 + 5 + 2 = 8$
    Sum of "plus" ranks : $8 + 4 + 10 + 9 + 6 + 7 + 3 = 47$

The smaller value of the two sums of ranks is the test statistic T. In this example $T = 8$.

5. Consult the table of the probability distribution of T (Appendix 6). When T is *equal to or less than* the critical value in the table, the null hypothesis is rejected at the particular level of significance. Enter the table at the appropriate value of N. N is not necessarily the total number of pairs of data, but the number of pairs less the number of pairs for which $d = 0$. Since there are none in the present example, $N = 10$. Entering Appendix 6 at $N = 10$, we find that our calculated value of T happens to be equal to the critical value of 8 at $P = 0.05$ for a two-tailed test. The difference is therefore statistically significant.

6. Record the result of the test as "there is a significant difference between the median weights of the two samples ($T = 8$, $P<0.05$, Wilcoxon's test for matched pairs)".

Table 14.3 Ranking of matched pairs

Sample A	Sample B	d	Rank of d
10.3	12.2	+1.9	+8
11.4	12.1	+0.7	+4
10.9	13.1	+2.2	+10
12.0	11.9	−0.1	−1
10.0	12.0	+2.0	+9
11.9	12.9	+1.0	+6
12.2	11.4	−0.8	−5
12.3	12.1	−0.2	−2
11.7	13.5	+1.8	+7
12.0	12.3	+0.3	+3

14.8 Restrictions and cautions in using the Wilcoxon's test for matched pairs

The Wilcoxon's test may only be applied when the value of one observation in a matched pair can be subtracted from the other. That is to say, they should be interval measurements, or counts of things. The number of matched pairs whose difference is *not* zero should be six or more. If the number of matched pairs exceeds about 40 the test is cumbersome and a parametric alternative (Section 14.13) is more appropriate.

1. Note that for H_0 to be rejected, T has to be *smaller* than or *equal* to the tabulated value at a given probability level.

2. The test is for a difference in *medians*. Do not make statements about sample *means*.

3. The test assumes that samples have been drawn from parent populations which are symmetrically but not necessarily normally distributed. Because it may be impossible to discern the shape of the distribution in small samples, the point is largely academic. Nevertheless, if it is known from other studies that the two samples have been drawn from populations which have fundamentally asymmetrical distributions, do not use the test.

14.9 Comparing means – parametric tests

Parametric tests that compare means are more restrictive than their non-parametric counterparts. First, data should be recorded on interval or ratio scales of measurement. Second, data should be approximately normally distributed; that they are so can be checked as described in Section 7.10. It is possible to transform certain data, for example counts of things and proportions, to normal by means of an appropriate transformation (see Chapter 8). A third restriction is that the populations from which the samples are drawn should have similar variances. The null hypothesis in a test for a difference between sample means is therefore:

H_0 :   two samples are drawn from populations with identical means and variances.

It follows that if the outcome of a test suggests the rejection of H_0, we should eliminate the possibility that it is due to a difference between *variances* rather than between means. There is a simple test (the F-test) which decides if the difference between two sample variances is so small that it may be ignored. This check should be applied routinely before testing the difference between means, for if the outcome of an *F*-test suggests that variances are not similar, then a test for difference between means cannot be validly applied.

14.10 The *F*-test (two-tailed)

Two populations which have identical variances also have, by definition, a variance ratio of unity (that is, 1.00). Small samples drawn from such populations have a variance ratio which is distributed by sampling error around unity. At which point a departure from unity is so great that it cannot be accounted for by sampling error (and the populations from the samples are drawn are presumed not to have equal variance) is decided by a variance ratio, or *F*-test. In that test, the null hypothesis is:

$$H_0: \quad \frac{\sigma_1^2}{\sigma_2^2} = F = 1$$

A significant departure from unity is checked in tables of the distribution of *F* at the appropriate degrees of freedom. Because the tabulated critical values of *F* are greater than 1, the greater variance is divided by the lesser variance:

$$F = \frac{\text{greater variance (Sample 1)}}{\text{lesser variance (Sample 2)}}$$

where the degrees of freedom (v) are $(n_1 - 1)$ and $(n_2 - 1)$ for Samples 1 and 2, respectively.

Example 14.4

The wing lengths of 25 male and 31 female Reed Buntings are measured and the means and standard deviations are determined for each sample. The variances, s^2, are obtained by squaring the standard deviations. All the necessary information is summarised below. It is convenient to designate the sample with the *larger* variance *Sample 1*.

Sample 1 Female	Sample 2 Male
$n_1 = 31$	$n_2 = 25$
$\bar{x}_1 = 78.2$	$\bar{x}_2 = 81.1$
$s_1 = 2.76$	$s_2 = 2.28$
$s_1^2 = 7.6$	$s_2^2 = 5.2$
$v_1 = (31-1) = 30$	$v_2 = (25-1) = 24$

Substituting in the formula for *F*,

$$F = \frac{\text{greater variance}}{\text{lesser variance}} = \frac{7.6}{5.2} = 1.46$$

Consulting the distribution of *F* at $P = 0.05$ in Appendix 7 (two-tailed test because before the samples are drawn it is not known which has the larger variance), we find the critical value at $v_1 = 30$ and $v_2 = 24$ is 2.21. Because our calculated value of 1.46 is less than this, we do not reject the null hypothesis. We conclude that the two samples have been drawn from populations with equal or very similar variances. We may now proceed with a test for a difference between means.

14.11 The z-test for comparing the means of two large samples

When two populations have identical means then $\mu_1 = \mu_2$ and the population mean difference $(\mu_1 - \mu_2) = 0$. The value of the difference between the means of two samples drawn from these populations can be expressed as a deviation from the population mean difference, namely $(\bar{x}_1 - \bar{x}_2) - 0$. When a deviation is divided by the *standard error of the difference* it is transformed to a *z-score* in the same way that a single observation is transformed to a z-score when its deviation from the mean is divided by the standard deviation (Section 7.7). Inserting the expression for standard error of the difference (Section 9.6), and noting that subtracting zero is inconsequential and need not be written in, the full expression becomes:

$$z = \frac{(\bar{x}_1 - \bar{x}_2)}{\sqrt{\dfrac{s_1^2}{n_1} + \dfrac{s_2^2}{n_2}}}$$

Thus, we can transform the mean difference of any two samples to a z-score. As we know from Section 7.7 the critical values of z are 1.96 and 2.58. When a calculated value of z exceeds these, H_0 is rejected at $P = 0.05$ or $P = 0.01$ respectively, and we conclude that the samples are unlikely to have been drawn from populations with identical means.

From the Central Limit Theorem (Section 9.2) we know that the means of a series of samples drawn from a single population are normally distributed. It follows then that in applying the z-test, the populations from which samples are drawn do not have to be normally distributed provided that the samples are quite large (over 30 observations). In cases where the populations are suspected to be badly skewed, the sample sizes should exceed 50.

Example 14.5

Substituting the relevant Reed Bunting data from Example 14.4 into the formula for the z-test we obtain:

$$z = \frac{\bar{x}_1 - \bar{x}_2}{\sqrt{\dfrac{s_1^2}{n_1} + \dfrac{s_2^2}{n_2}}} = \frac{78.2 - 81.1}{\sqrt{\dfrac{7.6}{31} + \dfrac{5.2}{25}}}$$

$$z = \frac{-2.9}{\sqrt{0.245 + 0.208}} = \frac{-2.9}{0.673} = -4.31$$

(The negative sign arises simply because, in this example, $\bar{x}_2$ is larger than $\bar{x}_1$; the sign is ignored in two-tailed tests).

Our calculated value of z exceeds the value of 2.58 which corresponds to $P = 0.01$. The difference between sample means is highly significant. We therefore reject H_0 in favour of the alternative hypothesis that samples are drawn from populations which have different means.

If there are independent grounds for predicting that one sample is drawn from a population which has a larger mean than the other, then a one-tailed test is called for. In a one-tailed test, the sample with the larger predicted mean is nominated Sample 1, in which case z will be positive only if H_1 is true. The critical value of z is lower in a one-tailed test, namely, 1.65 at $P = 0.05$. We reiterate our earlier recommendation. Use the more stringent two-tailed test unless there is a clear cut case for doing otherwise.

Readers should note that the value of 0.673 in the final step in the calculation of z above is the *standard error of the difference*. This is itself a useful number and its application is described in Section 9.6.

14.12 The t-test for comparing the means of two small samples

When samples are small (under about 30 observations in each) a different version of the z-test, called the **t-test** is called for. The rationale is the same, namely that the mean difference between samples is divided by the *standard error of the difference*. The answer is compared with the distribution of t at the appropriate degrees of freedom. As we explained in Section 9.7, the calculation of the standard error of the difference is a little more complicated in the case of small samples. Although the full formula for t appears rather cumbersome, all the terms within it are familiar, being the sample sizes, means and standard deviations:

$$t = \frac{(\bar{x}_1 - \bar{x}_2)}{\sqrt{\left[\dfrac{(n_1 - 1)s_1^2 + (n_2 - 1)s_2^2}{(n_1 + n_2 - 2)}\right]\left[\dfrac{n_1 + n_2}{n_1 n_2}\right]}}$$

where the degrees of freedom are $(n_1 + n_2) - 2$. Unlike the z test, the t test *does* assume that samples have been drawn from populations which are normally distributed. Like the z-test it also assumes that variances are approximately equal. This is tested by the F-test (Section 14.10).

Example 14.6

A summary of sample data for measurements of males and females is given below. Is it likely that they are drawn from populations with equal means; that is to say, are the sample means statistically significantly different?

Male (sample 1)	Female (sample 2)
$n_1 = 6$	$n_2 = 8$
$\bar{x}_1 = 74.8$	$\bar{x}_2 = 72.99$
$s_1 = 1.04$	$s_2 = 1.48$
$s_1^2 = 1.08$	$s_2^2 = 2.20$

First, a preliminary check to establish that variances are similar. Proceeding as in Example 14.4, we determine F to be $2.20/1.08 = 2.04$. Checking the distribution of F at $P = 0.05$ (two-tailed test) in Appendix 7, the tabulated value of F at 7 and 5 df respectively is 6.85. The calculated value of F is well below this and we proceed with the t test. Substituting in the formula for t:

$$t = \frac{(74.8 - 72.99)}{\sqrt{\left[\dfrac{(6 - 1)1.08 + (8 - 1)2.20}{(6 + 8 - 2)}\right]\left[\dfrac{6 + 8}{6 \times 8}\right]}}$$

$$= \frac{1.81}{\sqrt{\left[\dfrac{5.4 + 15.4}{12}\right] \times 0.2917}}$$

$$= \frac{1.81}{\sqrt{1.733 \times 0.2917}} = \frac{1.81}{0.711} = 2.55$$

(Note that the value of 0.711 in the final step is the *standard error of the difference*).

Consulting the table of the distribution of t (Appendix 1) we find that the calculated value of 2.55 exceeds the tabulated value of 2.179 at $P = 0.05$ for $(14 - 2) = 12$ df. We reject the null hypothesis and conclude there is a statistically significant difference between the means.

The t test may be applied as a one-tailed test exactly as described for the z test. The distribution of t in a one-tailed test is given in Appendix 1.

14.13 The t test for matched pairs

A specific form of the t test is used when measurements constitute a matched pair. As noted in Section 14.2, it is possible to conduct a more sensitive test upon matched data. The rationale for the t test with matched data is similar to that for the z and t tests with unmatched data. As we explain in Section 14.7 in connection with the Wilcoxon's test for matched pairs, it is possible to derive a difference, d, by subtracting the value of one observation in a matched pair from the other. If the null hypothesis that samples are drawn from populations with equal means is true, we would expect the mean value of d to be zero. We would not expect *all* values of d to be zero; rather, that all values would be normally distributed about the mean of zero. The standard deviation of the distribution is called the standard error of d. t is calculated by dividing the mean of the sample differences by the standard error of d. The overall formula for t is:

$$t = \frac{\Sigma d}{\sqrt{\frac{n\Sigma d^2 - (\Sigma d)^2}{(n-1)}}}$$

where n is the number of matched pairs and $(n-1)$ is the number of degrees of freedom.

Example 14.7

The Reed Warbler weight data employed to illustrate the Wilcoxon's test for matched pairs (Section 14.7) is used to demonstrate the procedure for computing the t test for matched pairs. We assume that the null hypothesis is that there is no difference between the means of the samples and that H_1 is that there is a difference, and calls for a two-tailed test. The table of data is reproduced in Table 14.4 and contains a column for the difference d between the values of each pair, and another for the squares of the difference (d^2). Values in these columns are then summed to give Σd and Σd^2 respectively. Notice that the sign is taken into account when summing the column and that the eventual value of t is not affected by which sample is nominated A or B.

From Σd we calculate $(\Sigma d)^2$ to be $8.8^2 = 77.44$. Substituting in the formula for t, with $n = 10$,

$$t = \frac{8.8}{\sqrt{\frac{(10 \times 17.96) - 77.44}{10 - 1}}} = \frac{8.8}{3.369} = 2.612$$

$t = 2.612$, at 9 degrees of freedom.

Consulting a table of the distribution of t (Appendix 1) we find that for a two-tailed test our calculated value of t at 9 df exceeds the tabulated value of 2.262 at $P = 0.05$. We therefore reject H_0 and conclude that the means of the two sets are significantly different in the two-tailed test.

Table 14.4. t test for matched pairs

Sample A Weight in August (g)	Sample B Weight in September (g)	d	d²
10.3	12.2	+1.9	3.61
11.4	12.1	+0.7	0.49
10.9	13.1	+2.2	4.84
12.0	11.9	−0.1	0.01
10.0	12.0	+2.0	4.0
11.9	12.9	+1.0	1.0
12.2	11.4	−0.8	0.64
12.3	12.1	−0.2	0.04
11.7	13.5	+1.8	3.24
12.0	12.3	+0.3	0.09
		$\Sigma d = 8.8$	$\Sigma d^2 = 17.96$

14.14 Restrictions and cautions when comparing means

1. If there are more than 30 observations in each sample use the z test. If the distribution of data appears to be badly skewed, increase the number of observations to 50. If this is not possible, try transforming the data or use the Mann-Whitney U test.

2. If there are fewer than 30 observations in each sample, use the t test. Unlike the z test, the t test *does* assume that data are derived from normally distributed populations. Badly skewed data can sometimes be 'normalised' by transformation (see Chapter 8). Conduct the test on the transformed observations but do not back-transform the final value of t. If the data defy all attempts to normalise them, use the Mann-Whitney U test.

3. Both z and t tests require that the variances of the two samples are similar. Check that they are so by means of a two-tailed F-test before conducting a z or t test.

4. z tests or t tests only answer the question: "Is there a statistically significant difference between the means of the two samples?" They do not address the more interesting question: "*To what extent* are the means different?" the execution of both z and t tests requires the intermediate calculation of the *standard error of the difference*. This statistic may be employed to estimate the difference between the means of the populations from which the samples are drawn (Section 9.6).

5. A test for the difference between the means of two samples can also be carried out with a *one-way analysis of variance*. This is the subject of Chapter 15.

CHAPTER 15
ANALYSIS OF VARIANCE – ANOVA

15.1 Why do we need ANOVA?

Chapter 14 discusses ways of comparing the means of two samples. Sometimes, however, ornithologists wish to compare the means of more than two samples. Suppose, for example, we have wing length measurements from samples of three races of a species each of which lives on different islands A, B and C. It is possible to compare the mean lengths by z tests or, if the samples are small, by t tests. We would need to perform the test three times to compare A–B, A–C and B–C. With the help of a calculator the task is not too daunting. Let us imagine instead that we wish to compare the means of seven samples. In this event no less than 21 z or t tests are required to compare all possible pairs of means. Even if the analyst has sufficient patience to work through such a cumbersome treatment, there is an underlying statistical objection to doing so.

We point out in Section 10.7 that if the $P = 0.05\%$ (5%) level of significance is consistently accepted, a wrong conclusion will be drawn on average once in every 20 tests performed. If the means of our hypothetical seven samples are compared in 21 z tests, there is a good chance that at least one false conclusion will be drawn. Of course the risk of committing a Type 1 error, that is, rejecting H_o when it should be accepted, is reduced by setting the acceptable significance level to the more stringent one of $P = 0.01$ (1%). But that increases the risk of making a Type 2 error, namely failing to reject H_o when it should be rejected. **Analysis of variance (ANOVA)** overcomes these difficulties by allowing comparisons to be made between any number of sample means, all in a single test. When it is used in this way to compare the means of several samples, statisticians speak of a *one-way ANOVA*.

ANOVA is such a flexible technique that it may also be used to compare more than one set of means. Referring to our island races again, it is possible to compare at the same time the mean lengths of samples of males and females of each race obtained from the islands. When the influence of two variables upon a sample mean is being analysed, such as island of origin and sex in our hypothetical example, the technique involved is described as a *two-way ANOVA*. Three-way, four-way (and so on) treatments are also possible but they get progressively more complicated. We restrict our examples in this guide to one-way and two-way treatments.

15.2 How ANOVA works

How analysis of *variance* is used to investigate differences between means is illustrated in the following example.

Example 15.1

Compare the individual variances of the three samples overleaf with the overall variance when all 15 observations ($n = 15$) are aggregated.

The means of Samples 1 and 2 are similar; the mean of Sample 3 is much lower; the mean of the aggregated observations is intermediate in value. The variances of the three samples are identical (10.00) and therefore the 'average variance' is 10.00. The variance of the aggregated observations however is larger (16.0) than the average sample variance. The increase is due to the difference between the *means* of the samples, in particular, the difference between the mean of Sample 3 and the other two means.

Sample 1	Sample 2	Sample 3	Overall
8	9	3	
10	11	5	
12	13	7	
14	15	9	
16	17	11	
$\Sigma x = 60$	$\Sigma x = 65$	$\Sigma x = 35$	$\Sigma x = 160$
$\bar{x} = 12.0$	$\bar{x} = 13.0$	$\bar{x} = 7.00$	$\bar{x} = 10.667$
$s^2 = 10.00$	$s^2 = 10.00$	$s^2 = 10.00$	$s^2 = 16.0$

The samples thus give rise to two sources of variability:

(i) the variability around the mean *within* a sample (random scatter);

(ii) the variability *between* the samples due to differences between the means of the populations from which the samples are drawn.

In other words:

$$\text{Variability}_{total} = \text{variability}_{within} + \text{variability}_{between}$$

ANOVA involves dividing up, or *partitioning*, the total variability of a number of samples into its components. If the samples are drawn from normally distributed populations with equal means and variances, the *within* variance is the same as the *between* variance. If a statistical test shows that this is not the case, then the samples have been drawn from populations with different means and/or variances. If it is assumed that the variances are equal (and this is an underlying assumption in ANOVA) then it is concluded that the discrepancy is due to differences between *means*. Thus:

H_0 = samples are drawn from normally distributed populations with equal means and variances.

H_1 = population variances are assumed to be equal and therefore samples are drawn from populations with different means.

As we shall explain, the assumption that population variances are equal is not to be taken for granted; there is a simple check that should be made before ANOVA is applied. If it is suspected that observations are not normally distributed (e.g. they are counts) then ANOVA is performed upon *transformed* observations (see Chapter 8).

When partitioning the total variability of a number of samples, it is simpler to work with *sums of squares* because adding and subtracting variances is complicated by varying numbers of degrees of freedom. However, in the final stages the sums of squares are converted to variances by dividing by the degrees of freedom in order to apply the *F* test to compare them.

In Section 6.6 we gave a formula for estimating sums of squares (SS) as:

$$SS = \Sigma x^2 - \frac{(\Sigma x)^2}{n}$$

The quantity $\dfrac{(\Sigma x)^2}{n}$ is often referred to as the **correction term** (CT).

15.3 Procedure for computing one-way ANOVA

An ornithologist wishes to know if the mean weights of Starlings sampled in four different roost situations are different. A sample of 10 units (Starlings) is obtained from each situation.
The procedure for conducting a one-way ANOVA is set out as a series of instructions.

1. Cast the data into a table, labelling each Sample 1 – 4, respectively. Use a scientific calculator to obtain, for each sample, the mean; the standard deviation (square this to obtain the variance); Σx (square this to obtain $(\Sigma x)^2$; and Σx^2. Record this information at the bottom of the column for each sample. At the right-hand side of the table record the sums of the totals of n, Σx, and Σx^2, using the subscript T to distinguish them from the sample data. These data are presented in Table 15.1.

Table 15.1. Weights of starlings from four roost situations (g)

Situation 1 Sample 1	Situation 2 Sample 2	Situation 3 Sample 3	Situation 4 Sample 4	Total
78	78	79	77	
88	78	73	69	
87	83	79	75	
88	81	75	70	
83	78	77	74	
82	81	78	83	
81	81	80	80	
80	82	78	75	
80	76	83	76	
89	76	84	75	
$n = 10$	$n = 10$	$n = 10$	$n = 10$	$n_T = 40$
$\bar{x} = 83.6$	$\bar{x} = 79.4$	$\bar{x} = 78.6$	$\bar{x} = 75.4$	
$s = 4.03$	$s = 2.50$	$s = 3.31$	$s = 4.14$	
$s^2 = 16.27$	$s^2 = 6.25$	$s^2 = 10.96$	$s^2 = 17.14$	
$\Sigma x = 836$	$\Sigma x = 794$	$\Sigma x = 786$	$\Sigma x = 754$	$\Sigma x_T = 3170$
$(\Sigma x)^2 = 698896$	$(\Sigma x)^2 = 630436$	$(\Sigma x)^2 = 617796$	$(\Sigma x)^2 = 568516$	
$\Sigma x^2 = 70036$	$\Sigma x^2 = 63100$	$\Sigma x^2 = 61878$	$\Sigma x^2 = 57006$	$\Sigma x_T^2 = 252020$

2. Before proceeding with the main part of the analysis it is necessary to check that all four sample variances are similar to each other. This is called a test for the **homogeneity of variance**; it is undertaken by means of the F_{max} test. Only one test is required. If the largest and smallest variances of the samples are not significantly different from each other, then the others cannot be. Select the largest variance in the table and divide it by the smallest. Equate the result to F:

$$F_{max} = \frac{17.14}{6.25} = 2.74, \text{ with 9 df in each sample}$$

Consulting a table of the distribution of F_{max} (Appendix 8) we find that our calculated value of F is less than the critical value of 6.31 for number of samples $a = 4$ and df $(n - 1) = 9$. We conclude that the variances are homogeneous and we proceed with ANOVA.

The next five steps involve the partitioning of the sums of squares into the categories which make up the total. In each case the subscripts 1–4 pertain to data from Samples 1–4 respectively.

3. Calculate a factor called the correction term, CT:

$$CT = \frac{(\Sigma x_T)^2}{n_T} = \frac{(3170)^2}{40} = 251222.5$$

4. Calculate the total sum of squares of the aggregated samples, SS_T:

$$SS_T = \Sigma x_T^2 - CT$$
$$SS_T = 252020 - 251222.5 = 797.5$$

5. Calculate the *between samples* sum of squares, $SS_{between}$

$$SS_{between} = \frac{(\Sigma x_1)^2}{n_1} + \frac{(\Sigma x_2)^2}{n_2} + \frac{(\Sigma x_3)^2}{n_3} + \frac{(\Sigma x_4)^2}{n_4} - CT$$

$$SS_{between} = \frac{698896}{10} + \frac{630436}{10} + \frac{617796}{10} + \frac{568516}{10} - CT$$

$$SS_{between} = 69889.6 + 63043.6 + 61779.6 + 56851.6 - 251222.5$$

$$SS_{between} = 341.9$$

6. We now need to know the *within samples* sum of squares. Since we already know (Section 15.2) that $SS_T = SS_{between} + SS_{within}$ we can derive SS_{within} simply by subtracting $SS_{between}$ from SS_T:

$$SS_{within} = (SS_T - SS_{between}) = (797.5 - 341.9) = 455.6$$

Whilst this 'short cut' method gives the correct result we advise SS_{within} be calculated independently because, if the sum of $SS_{between}$ and SS_{within} then checks with SS_T we can be assured that no mistake has been made in the calculation. SS_{within} is the sum of the individual SS_{within} for each sample. We designate the individual values as SS_W.

$$SS_{W1} = \Sigma x_1^2 - \frac{(\Sigma x_1)^2}{n_1} = 70036 - \frac{698896}{10} = 146.4$$

$$SS_{W2} = \Sigma x_2^2 - \frac{(\Sigma x_2)^2}{n_2} = 63100 - \frac{630436}{10} = 56.4$$

$$SS_{W3} = \Sigma x_3^2 - \frac{(\Sigma x_3)^2}{n_3} = 61878 - \frac{617796}{10} = 98.4$$

$$SS_{W4} = \Sigma x_4^2 - \frac{(\Sigma x_4)^2}{n_4} = 57006 - \frac{568516}{10} = 154.4$$

$$SS_{within} = 146.4 + 56.4 + 98.4 + 154.4 = 455.6$$

7. Check that the independently calculated values of SS_{within} and $SS_{between}$ add up to that of SS_T calculated in Step 4:

$$455.6 + 341.9 = 797.5$$

8. Determine the number of degrees of freedom (df) for each of the calculated SS values. The rules for determining these are:

df for $SS_T = n_T - 1 = 40 - 1 = 39$
df for $SS_{between} = a - 1$ (where a = number of samples) $= (4 - 1) = 3$
df for $SS_{within} = n_T - a = 40 - 4 = 36$

9. Estimate the variances by dividing each sum of squares by its respective degrees of freedom:

$$s^2_{between} = \frac{SS_{between}}{df_{between}} = \frac{341.9}{3} = 113.97$$

$$s^2_{within} = \frac{SS_{within}}{df_{within}} = \frac{455.6}{36} = 12.66$$

10. Compute F:

$$F = \frac{\text{Between samples variance}}{\text{Within samples variance}} = \frac{113.97}{12.66} = 9.002$$

It should be noted that the denominator, the bottom line in the division, is always the *within* samples variance. If it should turn out that this is larger than the *between* samples variance, then F is less than 1.0. This cannot be significant because tabulated values of F are greater than 1.0. Thus, there is no need to compute F. H_o is automatically accepted. Because a *nominated* variance (the *within* variance) is the denominator, this F test is *one-tailed* and the table in Appendix 9 is used.

11. Enter the result in an ANOVA summary table.

Source of variation	SS	df	s^2	F
Between	341.9	3	113.97	9.002
Within	455.6	36	12.66	
Total	797.5	39		

Consulting a table of the one-tailed distribution of F (Appendix 9), we find that our calculated value of F at 3 and 36 degrees of freedom exceeds the critical value of 2.88 (interpolating between 30 and 40 df in v_2). We therefore reject the null hypothesis, and conclude that the difference in the mean weight of the four Starling samples is significantly significant. We record the results as:

'The difference in mean weight of the four samples, where $n = 10$ in each case, is statistically significant ($F_{3,36} = 9.002$, $P < 0.05$).' The subscript 3,36 indicates the 3 and 36 degrees of freedom of the 'between' and 'within' variances, respectively.

This is not necessarily the end of the analysis, however. Taking the four means as a group, we know that there is a statistically significant variation between them. This may mean that all possible pairs are different from each other, or that just one is different from the other three. A good indication of which sample means are different from the others is obtained by presenting the individual sample means in histogram form, displaying the 95% confidence intervals, exactly as in Fig. 9.1. The samples whose intervals do not overlap are presumed to have been drawn from populations with different means. We suggest you always do this because it indicates whether the outcome of your ANOVA is 'reasonable'.

A more sensitive test for distinguishing the mean differences which are significantly different is the **Tukey Test**. This is simple to apply and is outlined in the next section.

15.4 Procedure for computing the Tukey Test

The Tukey test only needs to be undertaken when the result of the final F test in the ANOVA indicates that there is a significant difference between the means of the groups. This version can only be used when there is an equal number of observations in all samples. The procedure for the test is as follows.

1. Construct a trellis for the comparison of all sample means. This is done below for the Starling data used in the previous section. Any negative signs are ignored.

Sample	2	3	4
Sample 1 $\bar{x}_1 = 83.6$	$(\bar{x}_1 - \bar{x}_2)$ 4.2	$(\bar{x}_1 - \bar{x}_3)$ 5.0	$(\bar{x}_1 - \bar{x}_4)$ 8.2
Sample 2 $\bar{x}_2 = 79.4$		$(\bar{x}_2 - \bar{x}_3)$ 0.8	$(\bar{x}_2 - \bar{x}_4)$ 4.0
Sample 3 $\bar{x}_3 = 78.6$			$(\bar{x}_3 - \bar{x}_4)$ 3.2
Sample 4 $\bar{x}_4 = 75.4$			

2. Compute a test statistic T to provide a standard against which values in the trellis will be tested.

$$T = (q) \times \sqrt{\frac{\text{within variance}}{n}}$$

where n = the number of sampling units in each sample. The *within variance* is obtained from the ANOVA summary table of Step 11 in the previous section. The value of q is found by consulting a table of the distribution of q (the Tukey Table, Appendix 10) for varying numbers of degrees of freedom, v. The respective values in this example are 4 and 36 (36 being the df of the within samples variance). Interpolating Appendix 10 at $a = 4$, $v = 36$, the tabulated value of q is about 3.82 (interpolating between 30 and 40 df).

The T statistic is calculated as:

$$T = 3.82 \times \sqrt{12.66/10} = 4.30$$

There are only two mean differences in the trellis which exceed this: 1 and 3, and 1 and 4. These means are therefore significantly different at $P = 0.05$.

The values of q in the Tukey table have been calculated to take into account that several comparisons may be made and that there is a cumulative risk of committing a type 1 error. There is, of course, an associated increase in the risk of committing a type 2 error. Of the two, the latter is the more acceptable since it errs on the conservative side.

15.5 Two-way ANOVA

Two-way ANOVA techniques allow us to estimate the effects of two independent variables on a dependent variable. For example, a dependent variable might be weight; one of the independent variables might be sex, and season the other. Such data can be displayed in a contingency table. Instead of the data consisting of single counts, or frequencies, they represent the means of a number of observations. Hypothetical measurements for mean weights of both sexes of a bird species weighed in autumn and spring are set out in the table below.

		Variable B, sex	
		Male	Female
Variable A, season	Autumn	Sample 1 57 g	Sample 2 53 g
	Spring	Sample 3 55 g	Sample 4 51 g

There are two particular questions that arise from the data:

(i) Is there a significant difference between the mean weights of males and females?

(ii) Is there a significant difference betwen the mean weights of birds in each season?

There are six combinations of pairs of means we need to compare in order to answer these questions: autumn males – autumn females; autumn males – spring males; autumn males – spring females; autumn females – spring males; autumn females – spring females; spring males – spring females. Whilst it is certainly possible to compare these by means of z or t tests, the same objections that are raised in Section 15.1 apply here also. There is, however, an additional reason why these tests are inadequate: z and t tests fail to reveal any effect due to **interaction** of the two variables.

The idea of interaction is explained graphically. In Fig. 15.1a four sample means are plotted with hypothetical confidence limits. Sample means of each sex are joined by lines which are roughly parallel. These indicate that the effects of the two variables (sex and season) are *additive*. That is to say, transition from autumn to spring has added (or, in this example, subtracted – subtraction is just negative addition!) an equal amount from the mean weight of each sex.

Alternatively, we may take the view that transition from female to male has added an equal amount to the mean weight of each season. There is therefore no interaction between sex and season in Fig. 15.1a.

In Figs. 15.1b and 15.1c the situation is different. In Fig. 15.1b transition from autumn to spring results in a proportionately greater change to the mean weight of females, and the variables sex and season are interactive. In Fig. 15.1c there is also interaction, but of a different kind. Transition from autumn to spring results in a proportionately lesser change to females than males.

If interaction is observed then a biological explanation may be sought. Hypothetically, Fig. 15.1a might represent the mean weight of non-breeding males and females in each season. In Fig. 15.1b we might suppose that the spring females were weighed just after egg laying when body reserves are depleted. Fig. 15.1c might reflect spring females which were weighed just prior to egg laying.

If interaction between variables is suspected, plot out the means as shown in Figs. 15.1 and 15.2.

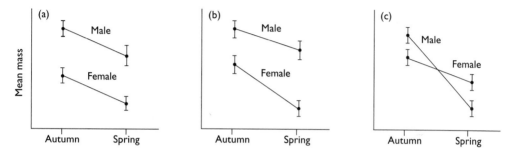

Fig. 15.1 Mean weights of samples of male and female birds weighed in autumn and spring: (a) no interaction; (b) positive interaction; (c) negative interaction. Confidence limits are hypothetical.

If there are more than two categories of each variable then the joined lines may form zig-zags. If the zig-zags are roughly parallel, there is no interaction; if the zig-zags are obviously not parallel then there may be interaction (Figs. 15.2a, b).

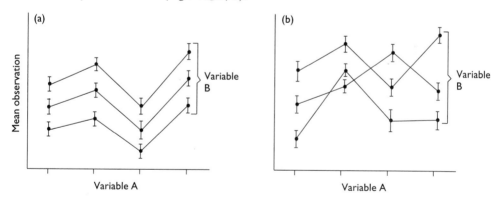

Fig. 15.2. Interactive effects between variables: (a) no interaction; (b) interaction. Confidence limits are hypothetical.

15.6 Partitioning the sums of squares in two-way ANOVA

Two-way ANOVA involves partitioning the total sums of squares of the samples into the various components that makes up the total variability. Once again there are two major components of the total sum of squares related in the equation:

$$SS_{total} = SS_{between} + SS_{within}$$

The SS_{within} quantity again represents the variability caused by random effects within each sample. In the two-way ANOVA, the $SS_{between}$ item is sub-divided into three components:

(i) SS representing variation between samples due to variable A (SS_A)

(ii) SS representing variation between samples due to variable B (SS_B)

(iii) SS representing variability between samples due to the interaction of variable A and variable B (SS_i).

The complete equation therefore may be written:

$$SS_T = (SS_A + SS_B + SS_i) + SS_{within}$$

In a two-way ANOVA we estimate all the components. The analysis concludes with separate comparisons of SS_A, SS_B and SS_i, with SS_{within}. As before, the individual values of SS are converted to variances by dividing by the appropriate degrees of freedom so the F tests may be used to check for significant differences between them.

15.7 Procedure for computing two-way ANOVA

Example 15.2

Suppose that the Starling data used to illustrate the one-way ANOVA procedure in Section 15.3 had been obtained in November. The ornithologist catches more birds in January in the same four roosting situations, and wishes to know not only if there are differences between the roosting situations, but also if there has been a change in weight over the winter. He also wishes to discover if there are any interactive effects between roosting situation and date of sampling. Two-way ANOVA is appropriate to the problem. The stepwise procedure is set out below.

1. Cast the measurements of each sample into a contingency table (in this case a 4 x 2 table), labelling each sample 1–8. Use a scientific calculator to obtain, for each sample, the mean; the standard deviation (square this to obtain s^2); Σx (square this to obtain $(\Sigma x)^2$; and Σx^2. Record this information with the sample data in each cell. At the side of each row, and at the foot of each column, record the sums of the total of n, Σx and Σx^2 for each row and column, respectively, using the subscript t to distinguish them from the sample data. In the bottom right hand cell of the table there is space to record the totals of these items for all rows (which is equal to the total for all columns) and represents the grand total. Use the subscript T to identify these. These data are presented in Table 15.2.

2. Homogeneity of variance check: select the largest sample variance in the table, divide it by the smallest, and equate the result to F_{max} (as shown in Section 15.3, step 2). In this example, F_{max} is computed as $29.92/10.24 = 2.92$, for $a = 8$ and df $= 9$. This is less than the tabulated critical value of 8.95 (Appendix 8) and so we proceed with ANOVA.

3. Calculate the correction term, CT:

$$CT = \frac{(\Sigma x_T)^2}{n_T} = \frac{(6704)^2}{80} = 561795.2$$

4. Calculate the total sum of squares of the aggregated samples, SS_T

$$SS_T = \Sigma x_T^2 - CT$$
$$SS_T = 565334 - 561795.2 = 3538.8$$

Table 15.2 Two-way ANOVA of Starling weights

VARIABLE B - ROOSTING SITUATION

VARIABLE A - MONTH

	Situation 1 Column 1	Situation 2 Column 2	Situation 3 Column 3	Situation 4 Column 4	Total for row
November Row 1	Sample 1 78 82 88 81 87 80 88 80 83 89 $n = 10$ $\bar{x} = 83.6$ $s^2 = 16.24$ $s = 4.03$ $\Sigma x = 836$ $(\Sigma x)^2 = 698896$ $\Sigma x^2 = 70036$	Sample 2 78 81 78 81 85 82 81 76 78 74 $n = 10$ $\bar{x} = 79.4$ $s^2 = 10.24$ $s = 3.20$ $\Sigma x = 794$ $(\Sigma x)^2 = 630436$ $\Sigma x^2 = 63136$	Sample 3 79 78 73 80 79 78 75 83 77 84 $n = 10$ $\bar{x} = 78.6$ $s^2 = 10.96$ $s = 3.31$ $\Sigma x = 786$ $(\Sigma x)^2 = 617796$ $\Sigma x^2 = 61878$	Sample 4 77 84 68 80 75 75 70 76 74 75 $n = 10$ $\bar{x} = 75.4$ $s^2 = 20.52$ $s = 4.53$ $\Sigma x = 754$ $(\Sigma x)^2 = 568516$ $\Sigma x^2 = 57036$	$n_t = 40$ $\Sigma x_t = 3170$ $\Sigma x_t^2 = 252086$
January Row 2	Sample 5 85 87 88 98 86 86 95 89 100 94 $n = 10$ $\bar{x} = 90.8$ $s^2 = 29.92$ $s = 5.47$ $\Sigma x = 908$ $\Sigma(x)^2 = 824464$ $\Sigma x^2 = 82716$	Sample 6 84 87 88 93 91 87 96 94 86 96 $n = 10$ $\bar{x} = 90.2$ $s^2 = 19.10$ $s = 4.37$ $\Sigma x = 902$ $\Sigma(x)^2 = 813604$ $\Sigma x^2 = 81532$	Sample 7 91 88 90 92 87 96 84 83 86 85 $n = 10$ $\bar{x} = 88.2$ $s^2 = 16.40$ $s = 4.05$ $\Sigma x = 882$ $(\Sigma x)^2 = 777924$ $\Sigma x^2 = 77940$	Sample 8 90 86 87 82 85 80 81 90 84 77 $n = 10$ $\bar{x} = 84.2$ $s^2 = 18.15$ $s = 4.26$ $\Sigma x = 842$ $(\Sigma x)^2 = 708964$ $\Sigma x^2 = 71060$	$n_t = 40$ $\Sigma x_t = 3534$ $\Sigma x_t^2 = 313248$
Total for column	$n_t = 20$ $\Sigma x_t = 1744$ $\Sigma x_t^2 = 152752$	$n_t = 20$ $\Sigma x_t = 1696$ $\Sigma x_t^2 = 144668$	$n_t = 20$ $\Sigma x_t = 1668$ $\Sigma x_t^2 = 139818$	$n_t = 20$ $\Sigma x_t = 1596$ $\Sigma x_t^2 = 128096$	$n_T = 80$ $\Sigma x_T = 6704$ $\Sigma x_T^2 = 565334$

5. Calculate the *between samples* sum of squares, $SS_{between}$

$$SS_{between} = \frac{(\Sigma x_1)^2}{n_1} + \frac{(\Sigma x_2)^2}{n_2} + \dots \frac{(\Sigma x_8)^2}{n_8} - CT$$

where the subscripts 1–8 pertain to data from samples 1–8 respectively.

$$SS_{between} = \frac{(836)^2}{10} + \frac{(794)^2}{10} + \frac{(786)^2}{10} + \frac{(754)^2}{10} + \frac{(908)^2}{10}$$

$$= \frac{(902)^2}{10} + \frac{(882)^2}{10} + \frac{(842)^2}{10} - CT$$

$$SS_{between} = 564060 - 561795.2$$

$$SS_{between} = 2264.8$$

6. Calculate the sum of squares for variable A.

$$SS_A = \frac{(\Sigma x_t \text{ row 1})^2}{n_t \text{ row 1}} + \frac{(\Sigma x_t \text{ row 2})^2}{n_t \text{ row 2}} - CT$$

$$SS_A = \frac{(3170)^2}{40} + \frac{(3534)^2}{40} - 561795.2$$

$$SS_A = 563451.4 - 561795.2$$

$$SS_A = 1656.2$$

7. Calculate the sum of squares for variable B, SS_B.

$$SS_B = \frac{(\Sigma x_t \text{ column 1})^2}{n_t \text{ column 1}} + \frac{(\Sigma x_t \text{ column 2})^2}{n_t \text{ column 2}} + \frac{(\Sigma x_t \text{ column 3})^2}{n_t \text{ column 3}} +$$

$$\frac{(\Sigma x_t \text{ column 4})^2}{n_t \text{ column 4}} - CT$$

$$SS_B = \frac{(1744)^2}{20} + \frac{(1696)^2}{20} + \frac{(1668)^2}{20} + \frac{(1596)^2}{20} - 561795.2$$

$$SS_b = 574.4$$

8. Calculate the sum of squares for the interaction, SS_i.

$$SS_i = SS_{between} - (SS_A + SS_B)$$

$$SS_i = 2264.8 - (1656.2 + 574.4)$$

$$SS_i = 34.2$$

9. Calculate the within sum of squares, SS within

$$SS_{within} = SS_T - SS_{between}$$

$$SS_{within} = 3538.8 - 2264.8$$

$$SS_{within} = 1274$$

Note that as in Section 15.3, step 6, it is possible to derive SS_{within} independently from SS_T and $SS_{between}$. This serves to confirm the accuracy of the calculations.

$$SS_{within} = \Sigma x_1^2 - \frac{(\Sigma x_1)^2}{n_1} + \ldots \Sigma x_8^2 - \frac{(\Sigma x_8)^2}{n_8}$$

10. Determine the degrees of freedom for each sum of squares. The rules are:

df for $SS_T = (n_T - 1) = 80 - 1 = 79$

df for $SS_{between} = (a - 1) = 8 - 1 = 7$ (where a = number of samples)

df for $SS_A = (r - 1) = 2 - 1 = 1$ (where r = number of rows)

df for $SS_B = (c - 1) = 4 - 1 = 3$ (where c = number of columns)

df for $SS_i = (r - 1)(c - 1) = 1 \times 3 = 3$

df for $SS_{within} = (n_T - rc) = 80 - (2 \times 4) = 72$

11. Estimate the variances by dividing all the sums of squares by their respective degrees of freedom.

$$Variance_T = \frac{SS_T}{df_T} = \frac{3538.8}{79} = 44.79$$

$$Variance_A = \frac{SS_A}{df_A} = \frac{1656.2}{1} = 1656.2$$

$$Variance_B = \frac{SS_B}{df_B} = \frac{574.4}{3} = 191.5$$

$$Variance_i = \frac{SS_i}{df_i} = \frac{34.2}{3} = 11.4$$

$$Variance_{within} = \frac{SS_{within}}{df_{within}} = \frac{1274}{72} = 17.69$$

12. Calculate the F values for the main effects.

$$F \text{ (variable A)} = \frac{Variance_A}{Variance_{within}} = \frac{1656.2}{17.69} = 93.62$$

$$F \text{ (variable B)} = \frac{Variance_B}{Variance_{within}} = \frac{191.5}{17.69} = 10.83$$

$$F \text{ (interaction)} = \frac{\text{Variance}_i}{\text{Variance}_{within}} = \frac{11.4}{17.69} = 0.64$$

13. Summarise the data in an ANOVA table:

Source of variation	Sum of squares	df	Variance	F
(Between samples)	(2264.8)	(7)		
Variable A	1656.2	1	1656.2	93.62**
Variable B	574.4	3	191.5	10.83**
Interaction	34.2	3	11.4	0.64
Within samples	1274	72	17.69	

** Significant at $P<0.01$.

14. Refer to a table of the distribution of F (Appendix 9). Two of our three F values exceed the critical value at $P= 0.01$ at the appropriate numbers of degrees of freedom. Note that because the within-samples variance is larger than the interaction variance there is no need to compute F for interaction.

Our first null hypothesis is that there are no significant differences between the weights of Starlings in different roosting situations (variable B). Our value of $F= 10.83$ exceeds that tabulated at $P = 0.01$, namely about 4.0 at df 3,72. We therefore reject H_o and conclude that roosting situation *does* affect the mean weight.

Our second null hypothesis is that there are no significant differences between the weights of Starlings captured on the two sampling dates. Our value of $F= 93.62$ greatly exceeds the tabulated value at $P = 0.01$ of approximately 6.9 for df 1,72. We therefore reject H_o and conclude that sampling date *does* affect the weight of Starlings.

Our third null hypothesis is that there is no interaction between roosting situation and sampling date which influences the mean weights. Our calculated value of $F= 0.64$ does not allow us to reject the null hypothesis. We therefore conclude that there is no interactive effect.

As is the case in our example of one-way ANOVA, this is not necessarily the end of the analysis. Although we know that there are statistically significant differences between the means of the samples, we do not know *which particular means* are significantly different. As we explained at the end of Section 15.3, plotting the means of each sample with their 95% confidence intervals will reveal obvious differences between samples. However, the Tukey Test is more sensitive for pin-pointing differences between means.

15.8 Procedure for computing the Tukey test in two-way ANOVA

The procedure is exactly the same as described in Section 15.4 for one-way ANOVA. Construct a trellis for the comparison of all sample means. Table 15.3 refers to the two-way ANOVA data from Table 15.2. Any negative signs are ignored.

Table 17.3. Tukey trellis for two-way ANOVA.

Sample	2	3	4	5	6	7	8
Sample 1 $\bar{x}=83.6$	$\bar{x}_1-\bar{x}_2$ 4.2	$\bar{x}_1-\bar{x}_3$ 5.0	$\bar{x}_1-\bar{x}_4$ 8.2	$\bar{x}_1-\bar{x}_5$ 7.2	$\bar{x}_1-\bar{x}_6$ 6.6	$\bar{x}_1-\bar{x}_7$ 4.6	$\bar{x}_1-\bar{x}_8$ 0.6
Sample 2 $\bar{x}=79.4$		$\bar{x}_2-\bar{x}_3$ 0.8	$\bar{x}_2-\bar{x}_4$ 4.0	$\bar{x}_2-\bar{x}_5$ 11.4	$\bar{x}_2-\bar{x}_6$ 10.8	$\bar{x}_2-\bar{x}_7$ 8.8	$\bar{x}_2-\bar{x}_8$ 4.8
Sample 3 $\bar{x}=78.6$			$\bar{x}_3-\bar{x}_4$ 3.2	$\bar{x}_3-\bar{x}_5$ 12.2	$\bar{x}_3-\bar{x}_6$ 11.6	$\bar{x}_3-\bar{x}_7$ 9.6	$\bar{x}_3-\bar{x}_8$ 5.6
Sample 4 $\bar{x}=75.4$				$\bar{x}_4-\bar{x}_5$ 15.4	$\bar{x}_4-\bar{x}_6$ 14.8	$\bar{x}_4-\bar{x}_7$ 12.8	$\bar{x}_4-\bar{x}_8$ 8.8
Sample 5 $\bar{x}=90.8$					$\bar{x}_5-\bar{x}_6$ 0.6	$\bar{x}_5-\bar{x}_7$ 2.6	$\bar{x}_5-\bar{x}_8$ 6.6
Sample 6 $\bar{x}=90.2$						$\bar{x}_6-\bar{x}_7$ 2.0	$\bar{x}_6-\bar{x}_8$ 6.0
Sample 7 $\bar{x}=88.2$							$\bar{x}_7-\bar{x}_8$ 4.0
Sample 8 $\bar{x}=84.2$							

Compute the test statistic from

$$T = (q)\sqrt{\frac{\text{within variance}}{n}}$$

where q is obtained from the Tukey Table (Appendix 10) as described in Section 15.4, and n is the number of observations in each sample. The values of a and v for these data are 8 and 72 respectively, giving us a tabulated value of q of approximately 4.4 (interpolating between 60 and 120 for v).

$$T = 4.4\sqrt{\frac{17.69}{10}} = 5.85$$

There are 15 out of a possible 28 pairs of means whose differences exceed this value and whose differences are, therefore, statistically significant.

15.9 Two-way ANOVA with single observations

In our example of two-way ANOVA described in Section 15.7 the data in each cell of the table consist of several (10) observations. It is not necessary to have several observations in each cell, but if there is only one observation we are unable to compute a *between samples* sum of squares (Step 5). Without this quantity it is impossible to derive a sum of squares for *interaction* and this source of variability cannot be investigated. In order to apply two-way ANOVA to single observations we are obliged to assume there is *no* interaction.

Example 15.3

An expedition wishes to compare the densities of Puffin burrows on four islands. Five 50 m x 50 m quadrats are marked out on each island and the number of burrows in each quadrat is counted.

To gain a representative coverage of the whole of each island, quadrats are deliberately placed somewhere on the north, south, east and west sides of each island, with the fifth in the centre. The *exact* position of each quadrat is determined randomly. It is possible to classify each count in two ways: according to *island* and according to *position on an island*. The data are presented in Table 15.4 and the numbers in bold are the number of burrows counted in each quadrat. Is there a variation in the number of burrows due to the island? Is there a significant effect due to position?

Before commencing the analysis, we should note two points. First, the five observations from each of the four islands could be assembled into four samples. We could test for a difference between the mean numbers of burrows per quadrat in each sample (island) by a Kruskal-Wallis test (Section 14.5) or a one-way ANOVA. But then possible effects due to position are ignored. Second, the data consist of counts. Bearing Section 8.1 in mind, we should transform them before proceeding with a parametric test. This has the effect of normalising the data, stabilising the variances and, in some circumstances, helping to reduce error due to interaction.

In this example is it easy to show that the variance of each "sample" is greater than the mean. Thus a logarithmic transformation is suitable. Moreover, because there are no zero counts, the simple logarithm of each count is sufficient. The logarithm of each count is shown in italic type in the Table 15.4. The actual counts are now ignored and the analysis is performed upon the transformed data. The respective means, s^2, Σx and Σx^2 are given at the end of each row and at the foot of each column. The bottom right cell gives the total for these, identified by the subscript T.

In this example the total sum of squares, SS_T can be partitioned into three categories:

$$SS_{Total} = SS_{Variable\ A} + SS_{Variable\ B} + SS_{Within}$$

As in our previous examples of ANOVA, we proceed as follows:

1. Calculate CT

$$CT = \frac{(\Sigma x_T)^2}{n_T} = \frac{(28.79)^2}{20} = 41.44$$

2. Calculate the total sum of squares, SS_T

$$SS_T = \Sigma x_T^2 - CT = 45.88 - 41.44 = 4.44$$

3. Calculate the sum of squares for VARIABLE A, SSA

$$SS_A = \frac{(\Sigma x_1)^2}{n_c} + \frac{(\Sigma x_2)^2}{n_c} + \frac{(\Sigma x_3)^2}{n_c} + \frac{(\Sigma x_4)^2}{n_c} - CT$$

Table 15.4 Two-way ANOVA on transformed counts. Numbers in italics are logarithms

VARIABLE B – POSITION / VARIABLE A – ISLAND

	Column 1 North	Column 2 South	Column 3 East	Column 4 West	Column 5 Centre	
Row 1 Bigga	96 *1.982*	24 *1.380*	70 *1.845*	195 *2.290*	62 *1.792*	$\bar{x} = 1.858$ $s^2 = 0.109$ $\Sigma x = 9.289$ $\Sigma x^2 = 17.69$
Row 2 Samphrey	12 *1.079*	14 *1.146*	45 *1.653*	39 *1.591*	50 *1.699*	$\bar{x} = 1.434$ $s^2 = 0.088$ $\Sigma x = 7.168$ $\Sigma x^2 = 10.63$
Row 3 Brother	3 *0.477*	7 *0.845*	14 *1.146*	4 *0.602*	19 *1.279*	$\bar{x} = 0.870$ $s^2 = 0.118$ $\Sigma x = 4.349$ $\Sigma x^2 = 4.253$
Row 4 Linga	84 *1.924*	10 *1.000*	51 *1.708*	32 *1.505*	71 *1.851*	$\bar{x} = 1.598$ $s^2 = 0.137$ $\Sigma x = 7.988$ $\Sigma x^2 = 13.31$
	$\bar{x} = 1.366$ $s^2 = 0.521$ $\Sigma x = 5.46$ $\Sigma x^2 = 9.02$	$\bar{x} = 1.093$ $s^2 = 0.052$ $\Sigma x = 4.371$ $\Sigma x^2 = 4.932$	$\bar{x} = 1.588$ $s^2 = 0.093$ $\Sigma x = 6.352$ $\Sigma x^2 = 10.367$	$\bar{x} = 1.497$ $s^2 = 0.480$ $\Sigma x = 5.988$ $\Sigma x^2 = 10.403$	$\bar{x} = 1.655$ $s^2 = 0.067$ $\Sigma x = 6.621$ $\Sigma x^2 = 11.16$	$n_T = 20$ $\Sigma x_T = 28.79$ $\Sigma x_T^2 = 45.88$

where n_c = number of columns (i.e. the number of observations in a row), and subscripts 1–4 pertain to rows 1–4.

$$SS_A = \frac{(9.289)^2}{5} + \frac{(7.168)^2}{5} + \frac{(4.349)^2}{5} + \frac{(7.988)^2}{5} - 41.44$$

$$SS_A = 44.08 - 41.44 = 2.64$$

4. Calculate the sum of squares for VARIABLE B, SS_B

$$SS_B = \frac{(\Sigma x_1)^2}{n_r} + \frac{(\Sigma x_2)^2}{n_r} + \frac{(\Sigma x_3)^2}{n_r} + \frac{(\Sigma x_4)^2}{n_r} + \frac{(\Sigma x_5)^2}{n_r} - CT$$

where n_r = number of rows (i.e. the number of observations in a column), and suffixes 1-5 pertain to columns 1–5.

$$SS_B = \frac{(5.46)^2}{4} + \frac{(4.371)^2}{4} + \frac{(6.352)^2}{4} + \frac{(5.988)^2}{4} + \frac{(6.621)^2}{4} - 41.44$$

$$SS_B = 42.24 - 41.44 = 0.800$$

5. Calculate the WITHIN sum of squares, SS_{within}

$$SS_{within} = SS_T - (SS_A + SS_B)$$

$$SS_{within} = 4.44 - (2.64 + 0.800) = 1.00$$

6. Determine the number of degrees of freedom for each sum of squares. The rules are:

$$\text{d.f. for } SS_T = n_T - 1 = 19$$
$$\text{d.f. for } SS_A = r - 1 = 3$$
$$\text{d.f. for } SS_B = c - 1 = 4$$
$$\text{d.f. for } SS_{within} = (r - 1)(c - 1) = 12$$

7. Estimate the variances by dividing each sum of squares by its respective degrees of freedom.

$$s^2_T = \frac{4.44}{19} = 0.234$$

$$s^2_A = \frac{2.64}{3} = 0.880$$

$$s^2_B = \frac{0.800}{4} = 0.200$$

$$s^2_{within} = \frac{1.00}{12} = 0.083$$

8. Calculate F for each effect

$$F\text{(variable A)} = \frac{Variance_A}{Variance_{within}} = \frac{0.880}{0.083} = 10.60$$

$$F\text{(variable B)} = \frac{Variance_B}{Variance_{within}} = \frac{0.200}{0.083} = 2.409$$

9. Summarise the results in an ANOVA table.

Source of variation	Sum of squares	df	s^2	F
Variable A	2.64	3	0.880	10.60**
Variable B	0.800	4	0.200	2.409
within	1.00	12	0.083	
Total	4.44	19		

*Significant at $P < 0.05$

Refer to a table of the probability distribution of F. Our value of 10.60 exceeds the tabulated value of 3.49 at 3,12 df ($P = 0.05$) in Appendix 9. Indeed, it exceeds the tabulated value of 5.9525 at $P=0.01$ (Appendix 9). We reject the null hypothesis and conclude that the difference in mean counts between the islands is statistically highly significant. The other value of 2.409 does not exceed the tabulated value of 3.26. We therefore accept H_0 and conclude that there does not appear to be any systematic variation between different parts of the islands.

This method may also be used for the analysis of measurements, in which case it may not be necessary to transform the data. It is particularly valuable for investigating variability between observers in recording counts or measurements (see *Ringing & Migration* 15: 84-90).

15.10 Analysis of variance in regression

We observe in Section 13.10 that analysis of variance may be used as an alternative to the *t* test for testing the significance of a *least squares* regression line. In Section 13.14 we note that ANOVA is the only suitable method for testing the significance of a regression line obtained by the r*educed major axis* method. The principle of ANOVA in regression is to determine the sum of squares due to the regression (*main effect*) and the residual sum of squares (*within*). Each is converted to a variance by dividing by the respective degrees of freedom, 1 and $(n-2)$, respectively. To complete the test the regression variance is divided by the residual variance to obtain an F value which is compared with the tabulated critical values of F at 1, $(n-2)$ df.

Example 15.4

Refer to the regression problem of Example 13.2. Establish the significance of the regression line by means of the F test in ANOVA. The quantities needed are derived in Section 13.8, page 93.

1. Calculate the regression sum of squares, $SS_{regression}$

$$SS_{regression} = \frac{(\text{sum of products})^2}{\text{sum of squares of } x}$$

These quantities are worked out in Steps (a) and (c) in Section 13.8.

$$SS_{regression} = \frac{(41837.5)^2}{51562.5} = 33946.694$$

The sum of squares for regression always has 1 df. Therefore the regression variance is equal to the regression sum of squares.

2. Calculate the residual variance. This quantity is obtained in Step (d) of Section 13.8 and is 360.98 with $(n-2)$ df where n is the number of pairs of observations. In this example there are $(10-2) = 8$ df.

3. Calculate F by dividing the regression variance by the residual variance:

$$F_{1,8} = \frac{33946.694}{360.98} = 94.040$$

4. Our calculated F value of 94.04 at 1,8 df greatly exceeds the tabulated critical value of 11.26 at these df at $P = 0.01$. We conclude that the regression is highly significant. The outcome is, of course, consistent with the result of the t test performed on the same data in Section 13.10.

15. Restrictions and cautions

ANOVA is an extremely efficient and powerful technique for investigating relationships between several groups of data. If the method at first seems difficult and complicated, it is certainly worth practising with data from your own field notebooks, following the steps we have laid out, until proficient.

1. ANOVA assumes that all observations are obtained randomly and are normally distributed. Certain data may be *transformed* to normal as described in Chapter 8.

2. ANOVA assumes that variances of the samples are similar. This can be checked with the F_{max} test. If the variances between samples are greatly different, transformation of data will usually stabilise the variances.

3. In performing the F test in ANOVA the *within* variance always appears in the denominator of the test. Because it is nominated as the smaller variance this F test is *one-tailed*. Tables of F distribution (Appendix 9) are published for a one-tailed test. Do not confuse them with the two-tailed tables (Appendix 7) using the equality of variance prior to a t test.

4. In one-way ANOVA, sample sizes do not have to be equal. In the example of two-way or interactive ANOVA we describe in Section 15.15, sample sizes must be equal. Field biologists

are often in a position to ensure that this is the case by deciding how many traps, quadrats, and so on, are to be set out. In Example 15.2, this is not the case, however. It is rather unlikely that equal numbers of Starlings will be captured in all situations. A practical expedient is to set the sample size to the smallest sample and select the same number of units to be measured randomly from the larger samples. This is admittedly wasteful, but the alternative is a complex analysis suitable for computer treatment and beyond the scope of this text.

5. In two-way ANOVA, interaction cannot be measured where data in each cell consist of single observations. Variability due to interaction is combined with the *within* variability and it is assumed to be negligible.

Appendix 1: t-distribution

df	Level of significance for one-tailed test			
	0.05	0.025	0.01	0.005
	Level of significance for two-tailed test			
	0.10	0.05	0.02	0.01
1	6.314	12.706	31.821	63.657
2	2.920	4.303	6.965	9.925
3	2.353	3.182	4.541	5.841
4	2.132	2.776	3.747	4.604
5	2.015	2.571	3.365	4.032
6	1.943	2.447	3.143	3.707
7	1.895	2.365	2.998	3.499
8	1.860	2.306	2.896	3.355
9	1.833	2.262	2.821	3.250
10	1.812	2.228	2.764	3.169
11	1.796	2.201	2.718	3.106
12	1.782	2.179	2.681	3.055
13	1.771	2.160	2.650	3.012
14	1.761	2.145	2.624	2.977
15	1.753	2.131	2.602	2.947
16	1.746	2.120	2.583	2.921
17	1.740	2.110	2.567	2.898
18	1.734	2.101	2.552	2.878
19	1.729	2.093	2.539	2.861
20	1.725	2.086	2.528	2.845
21	1.721	2.080	2.518	2.831
22	1.717	2.074	2.508	2.819
23	1.714	2.069	2.500	2.807
24	1.711	2.064	2.492	2.797
25	1.708	2.060	2.485	2.787
26	1.706	2.056	2.479	2.779
27	1.703	2.052	2.473	2.771
28	1.701	2.048	2.467	2.763
29	1.699	2.045	2.462	2.756
30	1.697	2.042	2.457	2.750
40	1.684	2.021	2.423	2.704
60	1.671	2.000	2.390	2.660
120	1.658	1.980	2.358	2.617
$\times$	1.645	1.960	2.326	2.576

Appendix 2: χ^2 distribution

Degrees of freedom	Level of significance	
	0.05	0.01
1	3.84	6.63
2	5.99	9.21
3	7.81	11.34
4	9.49	13.28
5	11.07	15.09
6	12.59	16.81
7	14.07	18.48
8	15.51	20.09
9	16.92	21.67
10	18.31	23.21
11	19.68	24.72
12	21.03	26.22
13	22.36	27.69
14	23.68	29.14
15	25.00	30.58
16	26.30	32.00
17	27.59	33.41
18	28.87	34.81
19	30.14	36.19
20	31.41	37.57
21	32.67	38.93
22	33.92	40.29
23	35.17	41.64
24	36.42	42.98
25	37.65	44.31
26	38.89	45.64
27	40.11	46.96
28	41.34	48.28
29	42.56	49.59
30	43.77	50.89
40	55.76	63.69
50	67.50	76.15
60	79.08	88.38
70	90.53	100.43
80	101.88	112.33
90	113.15	124.12
100	124.34	135.81

Appendix 3: Product moment correlation values at the 0.05 and 0.01 levels of significance

df	0.05	0.01	df	0.05	0.01
1	0.997	0.9999	32	0.339	0.436
2	0.950	0.990	34	0.329	0.424
3	0.878	0.959	35	0.325	0.418
4	0.811	0.917	36	0.320	0.413
5	0.754	0.874	38	0.312	0.403
6	0.707	0.834	40	0.304	0.393
7	0.666	0.798	42	0.297	0.384
8	0.632	0.765	44	0.291	0.376
9	0.602	0.735	45	0.288	0.372
10	0.576	0.708	46	0.284	0.368
11	0.553	0.684	48	0.279	0.361
12	0.532	0.661	50	0.273	0.354
13	0.514	0.641	55	0.261	0.338
14	0.497	0.623	60	0.250	0.325
15	0.482	0.606	65	0.241	0.313
16	0.468	0.590	70	0.232	0.302
17	0.456	0.575	75	0.224	0.292
18	0.444	0.561	80	0.217	0.283
19	0.433	0.549	85	0.211	0.275
20	0.423	0.537	90	0.205	0.267
21	0.413	0.526	95	0.200	0.260
22	0.404	0.515	100	0.195	0.254
23	0.396	0.505	125	0.174	0.228
24	0.388	0.496	150	0.159	0.208
25	0.381	0.487	175	0.148	0.193
26	0.374	0.479	200	0.138	0.181
27	0.367	0.471	300	0.113	0.148
28	0.361	0.463	400	0.098	0.128
29	0.355	0.456	500	0.088	0.115
30	0.349	0.449	1000	0.062	0.081

Appendix 4: Critical values of Spearman's Rank Correlation Coefficient

	Level of significance for one-tailed test			
	0.05	0.025	0.01	0.005
	Level of significance for two-tailed test			
n	0.10	0.05	0.02	0.01
5	0.900	–	–	–
6	0.829	0.886	0.943	–
7	0.714	0.786	0.893	–
8	0.643	0.738	0.833	0.881
9	0.600	0.683	0.783	0.833
10	0.564	0.648	0.745	0.794
11	0.523	0.623	0.736	0.818
12	0.497	0.591	0.703	0.780
13	0.475	0.566	0.673	0.745
14	0.457	0.545	0.646	0.716
15	0.441	0.525	0.623	0.689
16	0.425	0.507	0.601	0.666
17	0.412	0.490	0.582	0.645
18	0.399	0.476	0.564	0.625
19	0.388	0.462	0.549	0.608
20	0.377	0.450	0.534	0.591
21	0.368	0.438	0.521	0.576
22	0.359	0.428	0.508	0.562
23	0.351	0.418	0.496	0.549
24	0.343	0.409	0.485	0.537
25	0.336	0.400	0.475	0.526
26	0.329	0.392	0.465	0.515
27	0.323	0.385	0.456	0.505
28	0.317	0.377	0.448	0.496
29	0.311	0.370	0.440	0.487
30	0.305	0.364	0.432	0.478

Appendix 5: Mann-Whitney U-test values (two-tailed test) $P = 0.05$

n_1 \ n_2	2	3	4	5	6	7	8	9	10	11	12	13	14	15	16	17	18	19	20
2							0	0	0	0	1	1	1	1	1	2	2	2	2
3				0	1	1	2	2	3	3	4	4	5	5	6	6	7	7	8
4			0	1	2	3	4	4	5	6	7	8	9	10	11	11	12	13	13
5		0	1	2	3	5	6	7	8	9	11	12	13	14	15	17	18	19	20
6		1	2	3	5	6	8	10	11	13	14	16	17	19	21	22	24	25	27
7		1	3	5	6	8	10	12	14	16	18	20	22	24	26	28	30	32	34
8	0	2	4	6	8	10	13	15	17	19	22	24	26	29	31	34	36	38	41
9	0	2	4	7	10	12	15	17	20	23	26	28	31	34	37	39	42	45	48
10	0	3	5	8	11	14	17	20	23	26	29	33	36	39	42	45	48	52	55
11	0	3	6	9	13	16	19	23	26	30	33	37	40	44	47	51	55	58	62
12	1	4	7	11	14	18	22	26	29	33	37	41	45	49	53	57	61	65	69
13	1	4	8	12	16	20	24	28	33	37	41	45	50	54	59	63	67	72	76
14	1	5	9	13	17	22	26	31	36	40	45	50	55	59	64	67	74	78	83
15	1	5	10	14	19	24	29	34	39	44	49	54	59	64	70	75	80	85	90
16	1	6	11	15	21	26	31	37	42	47	53	59	64	70	75	81	86	92	98
17	2	6	11	17	22	28	34	39	45	51	57	63	67	75	81	87	93	99	105
18	2	7	12	18	24	30	36	42	48	55	61	67	74	80	86	93	99	106	112
19	2	7	13	19	25	32	38	45	52	58	65	72	78	85	92	99	106	113	119
20	2	8	13	20	27	34	41	48	55	62	69	76	83	90	98	105	112	119	127

n_1 and n_2 are the number of observations in each sample

Appendix 6: Critical values of T in the Wilcoxon's Test for two matched samples

	Levels of significance			
	One-tailed test			
	0.05	0.025	0.01	0.001
	Two-tailed test			
Sample size	0.1	0.05	0.02	0.002
N = 5	$T \leq 0$			
6	2	0		
7	3	2	0	
8	5	3	1	
9	8	5	3	
10	10	8	5	0
11	13	10	7	1
12	17	13	9	2
13	21	17	12	4
14	25	21	15	6
15	30	25	19	8
16	35	29	23	11
17	41	34	27	14
18	47	40	32	18
19	53	46	37	21
20	60	52	43	26
21	67	58	49	30
22	75	65	55	35
23	83	73	62	40
24	91	81	69	45
25	100	89	76	51
26	110	98	84	58
27	119	107	92	64
28	130	116	101	71
30	151	137	120	86
31	163	147	130	94
32	175	159	140	103
33	187	170	151	112

Appendix 7: F-distribution, 0.05 level of significance, two-tailed test

ν_2 \ ν_1	1	2	3	4	5	6	7	8	9	10	12	15	20	24	30	40	60	120	∞
1	647.8	799.5	864.2	899.6	921.8	937.1	948.2	956.7	963.3	968.6	976.7	984.9	993.1	997.2	1001	1006	1010	1014	1016
2	38.51	39.00	39.17	39.25	39.30	39.33	39.36	39.37	39.39	39.39	39.41	39.43	39.45	39.46	39.46	39.47	39.48	39.49	39.50
3	17.44	16.04	15.44	15.10	14.88	14.73	14.62	14.54	14.47	14.42	14.34	14.25	14.17	14.12	14.08	14.04	13.99	13.95	13.90
4	12.22	10.65	9.98	9.60	9.36	9.20	9.07	8.98	8.90	8.84	8.75	8.66	8.56	8.51	8.46	8.41	8.36	8.31	8.26
5	10.01	8.43	7.76	7.39	7.15	6.98	6.85	6.76	6.68	6.62	6.52	6.43	6.33	6.28	6.23	6.18	6.12	6.07	6.02
6	8.81	7.26	6.60	6.23	5.99	5.82	5.70	5.60	5.52	5.46	5.37	5.27	5.17	5.12	5.07	5.01	4.96	4.90	4.85
7	8.07	6.54	5.89	5.52	5.29	5.12	4.99	4.90	4.82	4.76	4.67	4.57	4.47	4.42	4.36	4.31	4.25	4.20	4.14
8	7.57	6.06	5.42	5.05	4.82	4.65	4.53	4.43	4.36	4.30	4.20	4.10	4.00	3.95	3.89	3.84	3.78	3.73	3.67
9	7.21	5.71	5.08	4.72	4.48	4.32	4.20	4.10	4.02	3.96	3.87	3.77	3.67	3.61	3.56	3.51	3.45	3.39	3.33
10	6.94	5.46	4.83	4.47	4.24	4.07	3.95	3.85	3.78	3.72	3.62	3.52	3.42	3.37	3.31	3.26	3.20	3.14	3.08
11	6.72	5.26	4.63	4.28	4.04	3.88	3.76	3.66	3.59	3.53	3.43	3.33	3.23	3.17	3.12	3.06	3.00	2.94	2.88
12	6.55	5.10	4.47	4.12	3.89	3.73	3.61	3.51	3.44	3.37	3.28	3.18	3.07	3.02	2.96	2.91	2.85	2.79	2.72
13	6.41	4.97	4.35	4.00	3.77	3.60	3.48	3.39	3.31	3.25	3.15	3.05	2.95	2.89	2.84	2.78	2.72	2.66	2.60
14	6.30	4.86	4.24	3.89	3.66	3.50	3.38	3.29	3.21	3.15	3.05	2.95	2.84	2.79	2.73	2.67	2.61	2.55	2.49
15	6.20	4.77	4.15	3.80	3.58	3.41	3.29	3.20	3.12	3.06	2.96	2.86	2.76	2.70	2.64	2.59	2.52	2.46	2.40
16	6.12	4.69	4.08	3.73	3.50	3.34	3.22	3.12	3.05	2.99	2.89	2.79	2.68	2.63	2.57	2.51	2.45	2.38	2.32
17	6.04	4.62	4.01	3.66	3.44	3.28	3.16	3.06	2.98	2.92	2.82	2.72	2.62	2.56	2.50	2.44	2.38	2.32	2.25
18	5.98	4.56	3.95	3.61	3.38	3.22	3.10	3.01	2.93	2.87	2.77	2.67	2.56	2.50	2.44	2.38	2.32	2.26	2.19
19	5.92	4.51	3.90	3.56	3.33	3.17	3.05	2.96	2.88	2.82	2.72	2.62	2.51	2.45	2.39	2.33	2.27	2.20	2.13
20	5.87	4.46	3.86	3.51	3.29	3.13	3.01	2.91	2.84	2.77	2.68	2.57	2.46	2.41	2.35	2.29	2.22	2.15	2.09
21	5.83	4.42	3.82	3.48	3.25	3.09	2.97	2.87	2.80	2.73	2.64	2.53	2.42	2.37	2.31	2.25	2.18	2.11	2.04
22	5.79	4.38	3.78	3.44	3.22	3.05	2.93	2.84	2.76	2.70	2.60	2.50	2.39	2.33	2.27	2.21	2.14	2.08	2.00
23	5.75	4.35	3.75	3.41	3.18	3.02	2.90	2.81	2.73	2.67	2.57	2.47	2.36	2.30	2.24	2.18	2.11	2.04	1.97
24	5.72	4.32	3.72	3.38	3.15	2.99	2.87	2.78	2.70	2.64	2.54	2.44	2.33	2.27	2.21	2.15	2.08	2.01	1.94
25	5.69	4.29	3.69	3.35	3.13	2.97	2.85	2.75	2.68	2.61	2.51	2.41	2.30	2.24	2.18	2.12	2.05	1.98	1.91
26	5.66	4.27	3.67	3.33	3.10	2.94	2.82	2.73	2.65	2.59	2.49	2.39	2.28	2.22	2.16	2.09	2.03	1.95	1.88
27	5.63	4.24	3.65	3.31	3.08	2.92	2.80	2.71	2.63	2.57	2.47	2.36	2.25	2.19	2.13	2.07	2.00	1.93	1.85
28	5.61	4.22	3.63	3.29	3.06	2.90	2.78	2.69	2.61	2.55	2.45	2.34	2.23	2.17	2.11	2.05	1.98	1.91	1.83
29	5.59	4.20	3.61	3.27	3.04	2.88	2.76	2.67	2.59	2.53	2.43	2.32	2.21	2.15	2.09	2.03	1.96	1.89	1.81
30	5.57	4.18	3.59	3.25	3.03	2.87	2.75	2.65	2.57	2.51	2.41	2.31	2.20	2.14	2.07	2.01	1.94	1.87	1.79
40	5.42	4.05	3.46	3.13	2.90	2.74	2.62	2.53	2.45	2.39	2.29	2.18	2.07	2.01	1.94	1.88	1.80	1.71	1.64
60	5.29	3.93	3.34	3.01	2.79	2.63	2.51	2.41	2.33	2.27	2.17	2.06	1.94	1.88	1.82	1.74	1.67	1.58	1.48
120	5.15	3.80	3.23	2.89	2.67	2.52	2.39	2.30	2.22	2.16	2.05	1.94	1.82	1.76	1.69	1.61	1.53	1.43	1.31
∞	5.02	3.69	3.12	2.79	2.57	2.41	2.29	2.19	2.11	2.05	1.94	1.83	1.71	1.64	1.57	1.48	1.39	1.27	1.00

Use this table for checking equality of variance prior to a z-test or a t-test

ν_1, ν_2 are the degrees of freedom of the samples with larger and smaller variances, respectively.

Appendix 8: Critical values of F_{max} 0.05 level of significance

Use this table when checking homogeneity of variance preceding Analysis of Variance

v	a	2	3	4	5	6	7	8	9	10	11	12
2		39.0	87.5	142	202	266	333	403	475	550	626	704
3		15.4	27.8	39.2	50.7	62.0	72.9	83.5	93.9	104	114	124
4		9.60	15.5	20.6	25.2	29.5	33.6	37.5	41.1	44.6	48.0	51.4
5		7.15	10.8	13.7	16.3	18.7	20.8	22.9	24.7	26.5	28.2	29.9
6		5.82	8.38	10.4	12.1	13.7	15.0	16.3	17.5	18.6	19.7	20.7
7		4.99	6.94	8.44	9.70	10.8	11.8	12.7	13.5	14.3	15.1	15.8
8		4.43	6.00	7.18	8.12	9.03	9.78	10.5	11.1	11.7	12.2	12.7
9		4.03	5.34	6.31	7.11	7.80	8.41	8.95	9.45	9.91	10.3	10.7
10		3.72	4.85	5.67	6.34	6.92	7.42	7.87	8.28	8.66	9.01	9.34
12		3.28	4.16	4.79	5.30	5.72	6.09	6.42	6.72	7.00	7.25	7.48
15		2.86	3.54	4.01	4.37	4.68	4.95	5.19	5.40	5.59	5.77	5.93
20		2.46	2.95	3.29	3.54	3.76	3.94	4.10	4.24	4.37	4.49	4.59
30		2.07	2.40	2.61	2.78	2.91	3.02	3.12	3.21	3.29	3.36	3.39
60		1.67	1.85	1.96	2.04	2.11	2.17	2.22	2.26	2.30	2.33	2.36

a is the number of samples being compared. v is the degrees of freedom of each sample (if samples do not have equal numbers of observations then use the degrees of freedom of the sample with the smaller number of observations).

Appendix 9: F - distribution (a) 0.05 level

Use these tables for testing significance in Analysis of Variance

v_1 = df for the greater variance v_2 = df for the lesser variance

$v_2 \backslash v_1$	1	2	3	4	5	6	7	8	9	10	12	15	20	24	30	40	60	120	∞
1	161.45	199.50	215.71	224.58	230.16	233.99	236.77	238.88	240.54	241.88	243.91	245.95	248.01	249.05	250.10	251.14	252.20	253.25	254.31
2	18.513	19.000	19.164	19.247	19.296	19.330	19.353	19.371	19.385	19.396	19.413	19.429	19.446	19.454	19.462	19.471	19.479	19.487	19.496
3	10.128	9.5521	9.2766	9.1172	9.0135	8.9406	8.8867	8.8452	8.8123	8.7855	8.7446	8.7029	8.6602	8.6385	8.6166	8.5944	8.5720	8.5594	8.5264
4	7.7086	6.9443	6.5914	6.3882	6.2561	6.1631	6.0942	6.0410	5.9938	5.9644	5.9117	5.8578	5.8025	5.7744	5.7459	5.7170	5.6877	5.6581	5.6281
5	6.6079	5.7861	5.4095	5.1922	5.0503	4.9503	4.8759	4.8183	4.7725	4.7351	4.6777	4.6188	4.5581	4.5272	4.4957	4.4638	4.4314	4.3985	4.3650
6	5.9874	5.1433	4.7571	4.5337	4.3874	4.2839	4.2067	4.1468	4.0990	4.0600	3.9999	3.9381	3.8742	3.8415	3.8082	3.7743	3.7398	3.7047	3.6689
7	5.5914	4.7174	4.3468	4.1203	3.9715	3.8660	3.7870	3.7257	3.6767	3.6365	3.5747	3.5107	3.4445	3.4105	3.3758	3.3404	3.3043	3.2674	3.2298
8	5.3177	4.4590	4.0662	3.8379	3.6875	3.5806	3.5005	3.4381	3.3881	3.3472	3.2839	3.2184	3.1503	3.1152	3.0794	3.0428	3.0053	2.9669	2.9276
9	5.1174	4.2565	3.8625	3.6331	3.4817	3.3738	3.2927	3.2296	3.1789	3.1373	3.0729	3.0061	2.9365	2.9005	2.8637	2.8259	2.7872	2.7475	2.7067
10	4.9646	4.1028	3.7083	3.4780	3.3258	3.2172	3.1355	3.0717	3.0204	2.9782	2.9130	2.8450	2.7740	2.7372	2.6996	2.6609	2.6211	2.5801	2.5379
11	4.8443	3.9823	3.5874	3.3567	3.2039	3.0946	3.0123	2.9480	2.8962	2.8536	2.7876	2.7186	2.6464	2.6090	2.5705	2.5309	2.4901	2.4480	2.4045
12	4.7472	3.8853	3.4903	3.2592	3.1059	2.9961	2.9134	2.8486	2.7964	2.7534	2.6866	2.6169	2.5436	2.5055	2.4663	2.4259	2.3842	2.3410	2.2962
13	4.6672	3.8056	3.4105	3.1791	3.0254	2.9153	2.8321	2.7669	2.7144	2.6710	2.6037	2.5331	2.4589	2.4202	2.3803	2.3392	2.2966	2.2524	2.2064
14	4.6001	3.7389	3.3439	3.1122	2.9582	2.8477	2.7642	2.6987	2.6458	2.6022	2.5342	2.4630	2.3879	2.3487	2.3082	2.2664	2.2229	2.1778	2.1307
15	4.5431	3.6823	3.2874	3.0556	2.9013	2.7905	2.7066	2.6408	2.5876	2.5437	2.4753	2.4034	2.3275	2.2878	2.2468	2.2043	2.1601	2.1141	2.0658
16	4.4940	3.6337	3.2389	3.0069	2.8524	2.7413	2.6572	2.5911	2.5377	2.4935	2.4247	2.3522	2.2756	2.2354	2.1938	2.1507	2.1058	2.0589	2.0096
17	4.4513	3.5915	3.1968	2.9647	2.8100	2.6987	2.6143	2.5480	2.4943	2.4499	2.3807	2.3077	2.2304	2.1898	2.1477	2.1040	2.0584	2.0107	1.9604
18	4.4139	3.5546	3.1599	2.9277	2.7729	2.6613	2.5767	2.5102	2.4563	2.4117	2.3421	2.2686	2.1906	2.1497	2.1071	2.0629	2.0166	1.9681	1.9168
19	4.3807	3.5219	3.1274	2.8951	2.7401	2.6283	2.5435	2.4768	2.4227	2.3779	2.3080	2.2341	2.1555	2.1141	2.0712	2.0264	1.9795	1.9302	1.8780
20	4.3512	3.4928	3.0984	2.8661	2.7109	2.5990	2.5140	2.4471	2.3928	2.3479	2.2776	2.2033	2.1242	2.0825	2.0391	1.9938	1.9464	1.8963	1.8432
21	4.3248	3.4668	3.0725	2.8401	2.6848	2.5727	2.4876	2.4205	2.3660	2.3210	2.2504	2.1757	2.0960	2.0540	2.0102	1.9645	1.9165	1.8657	1.8117
22	4.3009	3.4434	3.0491	2.8167	2.6613	2.5491	2.4638	2.3965	2.3419	2.2967	2.2258	2.1508	2.0707	2.0283	1.9842	1.9380	1.8894	1.8380	1.7831
23	4.2793	3.4221	3.0280	2.7955	2.6400	2.5277	2.4422	2.3748	2.3201	2.2747	2.2036	2.1282	2.0476	2.0050	1.9605	1.9139	1.8648	1.8128	1.7570
24	4.2597	3.4028	3.0088	2.7763	2.6207	2.5082	2.4226	2.3551	2.3002	2.2547	2.1834	2.1077	2.0267	1.9838	1.9390	1.8920	1.8424	1.7896	1.7330
25	4.2417	3.3852	2.9912	2.7587	2.6030	2.4904	2.4047	2.3371	2.2821	2.2365	2.1649	2.0889	2.0075	1.9643	1.9192	1.8718	1.8217	1.7684	1.7110
26	4.2252	3.3690	2.9752	2.7426	2.5868	2.4741	2.3883	2.3205	2.2655	2.2197	2.1479	2.0716	1.9898	1.9464	1.9010	1.8533	1.8027	1.7488	1.6906
27	4.2100	3.3541	2.9604	2.7278	2.5719	2.4591	2.3732	2.3053	2.2501	2.2043	2.1323	2.0558	1.9736	1.9299	1.8842	1.8361	1.7851	1.7306	1.6717
28	4.1960	3.3404	2.9467	2.7141	2.5581	2.4453	2.3593	2.2913	2.2360	2.1900	2.1179	2.0411	1.9586	1.9147	1.8687	1.8203	1.7689	1.7138	1.6541
29	4.1830	3.3277	2.9340	2.7014	2.5454	2.4324	2.3463	2.2783	2.2229	2.1768	2.1045	2.0275	1.9446	1.9005	1.8543	1.8055	1.7537	1.6981	1.6376
30	4.1709	3.3158	2.9223	2.6896	2.5336	2.4205	2.3343	2.2662	2.2107	2.1646	2.0921	2.0148	1.9317	1.8874	1.8409	1.7918	1.7396	1.6835	1.6223
40	4.0847	3.2317	2.8387	2.6060	2.4495	2.3359	2.2490	2.1802	2.1240	2.0772	2.0035	1.9245	1.8389	1.7929	1.7444	1.6928	1.6373	1.5766	1.5089
60	4.0012	3.1504	2.7581	2.5252	2.3683	2.2541	2.1665	2.0970	2.0401	1.9926	1.9174	1.8364	1.7480	1.7001	1.6491	1.5943	1.5343	1.4673	1.3893
120	3.9201	3.0718	2.6802	2.4472	2.2899	2.1750	2.0868	2.0164	1.9588	1.9105	1.8337	1.7505	1.6587	1.6084	1.5543	1.4952	1.4290	1.3519	1.2539
∞	3.8415	2.9957	2.6049	2.3719	2.2141	2.0986	2.0096	1.9384	1.8799	1.8307	1.7522	1.6664	1.5705	1.5173	1.4591	1.3940	1.3180	1.2214	1.0000

Appendix 9 (cont): F - distribution (b) 0.01 level

v_2 \ v_1	1	2	3	4	5	6	7	8	9	10	12	15	20	24	30	40	60	120	∞
1	4052.2	4999.5	5403.4	5624.6	5763.6	5859.6	5928.4	5981.1	6022.5	6055.8	6106.3	6157.3	6208.7	6234.6	6260.6	6286.6	6313.0	6339.4	6365.9
2	98.503	99.000	99.166	99.249	99.299	99.333	99.356	99.374	99.388	99.399	99.416	99.433	99.449	99.458	99.466	99.474	99.482	99.491	99.499
3	34.116	30.817	29.457	28.710	28.237	27.911	27.672	27.489	27.345	27.229	27.052	26.872	26.690	26.598	26.505	26.411	26.316	26.221	26.125
4	21.198	18.000	16.694	15.977	15.522	15.207	14.976	14.799	14.659	14.546	14.374	14.198	14.020	13.929	13.838	13.745	13.652	13.558	13.463
5	16.258	13.274	12.060	11.392	10.967	10.672	10.456	10.289	10.158	10.051	9.8883	9.7222	9.5526	9.4665	9.3793	9.2912	9.2020	9.1118	9.0204
6	13.745	10.925	9.7795	9.1483	8.7459	8.4661	8.2600	8.1017	7.9761	7.8741	7.7183	7.5590	7.3958	7.3127	7.2285	7.1432	7.0567	6.9690	6.8800
7	12.246	9.5466	8.4513	7.8466	7.4604	7.1914	6.9928	6.8400	6.7188	6.6201	6.4691	6.3143	6.1554	6.0743	5.9920	5.9084	5.8236	5.7373	5.6495
8	11.259	8.6491	7.5910	7.0061	6.6318	6.3707	6.1776	6.0289	5.9106	5.8143	5.6667	5.5151	5.3591	5.2793	5.1981	5.1156	5.0316	4.9461	4.8588
9	10.561	8.0215	6.9919	6.4221	6.0569	5.8018	5.6129	5.4671	5.3511	5.2565	5.1114	4.9621	4.8080	4.7290	4.6486	4.5666	4.4831	4.3978	4.3105
10	10.044	7.5594	6.5523	5.9943	5.6363	5.3858	5.2001	5.0567	4.9424	4.8491	4.7059	4.5581	4.4054	4.3269	4.2469	4.1653	4.0819	3.9965	3.9090
11	9.6460	7.2057	6.2167	5.6683	5.3160	5.0692	4.8861	4.7445	4.6315	4.5393	4.3974	4.2509	4.0990	4.0209	3.9411	3.8596	3.7761	3.6904	3.6024
12	9.3302	6.9266	5.9525	5.4120	5.0643	4.8206	4.6395	4.4994	4.3875	4.2961	4.1553	4.0096	3.8584	3.7805	3.7008	3.6192	3.5355	3.4494	3.3608
13	9.0738	6.7010	5.7394	5.2053	4.8616	4.6204	4.4410	4.3021	4.1911	4.1003	3.9603	3.8154	3.6646	3.5868	3.5070	3.4253	3.3413	3.2548	3.1654
14	8.8616	6.5149	5.5639	5.0354	4.6950	4.4558	4.2779	4.1399	4.0297	3.9394	3.8001	3.6567	3.5052	3.4274	3.3476	3.2656	3.1813	3.0942	3.0040
15	8.6831	6.3589	5.4170	4.8932	4.5556	4.3183	4.1415	4.0045	3.8948	3.8049	3.6662	3.5222	3.3719	3.2940	3.2141	3.1319	3.0471	2.9595	2.8684
16	8.5310	6.2262	5.2922	4.7726	4.4374	4.2016	4.0259	3.8896	3.7804	3.6909	3.5527	3.4089	3.2587	3.1808	3.1007	3.0182	2.9330	2.8447	2.7528
17	8.3997	6.1121	5.1850	4.6690	4.3359	4.1015	3.9267	3.7910	3.6822	3.5931	3.4552	3.3117	3.1615	3.0835	3.0032	2.9205	2.8348	2.7459	2.6530
18	8.2854	6.0129	5.0919	4.5790	4.2479	4.0146	3.8406	3.7054	3.5971	3.5082	3.3706	3.2273	3.0771	2.9990	2.9185	2.8354	2.7493	2.6597	2.5660
19	8.1849	5.9259	5.0103	4.5003	4.1708	3.9386	3.7653	3.6305	3.5225	3.4338	3.2965	3.1533	3.0031	2.9249	2.8442	2.7608	2.6742	2.5839	2.4893
20	8.0960	5.8489	4.9382	4.4307	4.1027	3.8714	3.6987	3.5644	3.4567	3.3682	3.2311	3.0880	2.9377	2.8594	2.7785	2.6947	2.6077	2.5168	2.4212
21	8.0166	5.7804	4.8740	4.3688	4.0421	3.8117	3.6396	3.5056	3.3981	3.3098	3.1730	3.0300	2.8796	2.8010	2.7200	2.6359	2.5484	2.4568	2.3603
22	7.9454	5.7190	4.8166	4.3134	3.9880	3.7583	3.5867	3.4530	3.3458	3.2576	3.1209	2.9779	2.8274	2.7488	2.6675	2.5831	2.4951	2.4029	2.3055
23	7.8811	5.6637	4.7649	4.2636	3.9392	3.7102	3.5390	3.4057	3.2986	3.2106	3.0740	2.9311	2.7805	2.7017	2.6202	2.5355	2.4471	2.3542	2.2558
24	7.8229	5.6136	4.7181	4.2184	3.8951	3.6667	3.4959	3.3629	3.2560	3.1681	3.0316	2.8887	2.7380	2.6591	2.5773	2.4923	2.4035	2.3100	2.2107
25	7.7698	5.5680	4.6755	4.1774	3.8550	3.6272	3.4568	3.3239	3.2172	3.1294	2.9931	2.8502	2.6993	2.6203	2.5383	2.4530	2.3637	2.2696	2.1694
26	7.7213	5.5263	4.6366	4.1400	3.8183	3.5911	3.4210	3.2884	3.1818	3.0941	2.9578	2.8150	2.6640	2.5848	2.5026	2.4170	2.3273	2.2325	2.1315
27	7.6767	5.4881	4.6009	4.1056	3.7848	3.5580	3.3882	3.2558	3.1494	3.0618	2.9256	2.7827	2.6316	2.5522	2.4699	2.3840	2.2938	2.1985	2.0965
28	7.6356	5.4529	4.5681	4.0740	3.7539	3.5276	3.3581	3.2259	3.1195	3.0320	2.8959	2.7530	2.6017	2.5223	2.4397	2.3535	2.2629	2.1670	2.0642
29	7.5977	5.4204	4.5378	4.0449	3.7254	3.4995	3.3303	3.1982	3.0920	3.0045	2.8685	2.7256	2.5742	2.4946	2.4118	2.3253	2.2344	2.1379	2.0342
30	7.5625	5.3903	4.5097	4.0179	3.6990	3.4735	3.3045	3.1726	3.0665	2.9791	2.8431	2.7002	2.5487	2.4689	2.3860	2.2992	2.2079	2.1108	2.0062
40	7.3141	5.1785	4.3126	3.8283	3.5138	3.2910	3.1238	2.9930	2.8876	2.8005	2.6648	2.5216	2.3689	2.2880	2.2034	2.1142	2.0194	1.9172	1.8047
60	7.0771	4.9774	4.1259	3.6490	3.3389	3.1187	2.9530	2.8233	2.7185	2.6318	2.4961	2.3523	2.1978	2.1154	2.0285	1.9360	1.8363	1.7263	1.6006
120	6.8509	4.7865	3.9491	3.4795	3.1735	2.9559	2.7918	2.6629	2.5586	2.4721	2.3363	2.1915	2.0346	1.9500	1.8600	1.7628	1.6557	1.5330	1.3805
∞	6.6349	4.6052	3.7816	3.3192	3.0173	2.8020	2.6393	2.5113	2.4073	2.3209	2.1847	2.0385	1.8783	1.7908	1.6964	1.5923	1.4730	1.3246	1.0000

Appendix 10: Tukey Test

($p = 0.05$)
a = the total number of means being compared
v = degrees of freedom of denominator of F test

v	a 2	3	4	5	6	7	8	9	10	11	12	13	14	15	16	17	18	19	20
1	17.97	26.98	32.82	37.08	40.41	43.12	45.40	47.36	49.07	50.59	51.96	53.20	54.33	55.36	56.32	57.22	58.04	58.83	59.56
2	6.08	8.33	9.80	10.88	11.74	12.44	13.03	13.54	13.99	14.39	14.75	15.08	15.38	15.65	15.91	16.14	16.37	16.57	16.77
3	4.50	5.91	6.82	7.50	8.04	8.48	8.85	9.18	9.46	9.72	9.95	10.15	10.35	10.52	10.69	10.84	10.98	11.11	11.24
4	3.93	5.04	5.76	6.29	6.71	7.05	7.35	7.60	7.83	8.03	8.21	8.37	8.52	8.66	8.79	8.91	9.03	9.13	9.23
5	3.64	4.60	5.22	5.67	6.03	6.33	6.58	6.80	6.99	7.17	7.32	7.47	7.60	7.72	7.83	7.93	8.03	8.12	8.21
6	3.46	4.34	4.90	5.30	5.63	5.90	6.12	6.32	6.49	6.65	6.79	6.92	7.03	7.14	7.24	7.34	7.43	7.51	7.59
7	3.34	4.16	4.68	5.06	5.36	5.61	5.82	6.00	6.16	6.30	6.43	6.55	6.66	6.76	6.85	6.94	7.02	7.10	7.17
8	3.26	4.04	4.53	4.89	5.17	5.40	5.60	5.77	5.92	6.05	6.18	6.29	6.39	6.48	6.57	6.65	6.73	6.80	6.87
9	3.20	3.95	4.41	4.76	5.02	5.24	5.43	5.59	5.74	5.87	5.98	6.09	6.19	6.28	6.36	6.44	6.51	6.58	6.64
10	3.15	3.88	4.33	4.65	4.91	5.12	5.30	5.46	5.60	5.72	5.83	5.93	6.03	6.11	6.19	6.27	6.34	6.40	6.47
11	3.11	3.82	4.26	4.57	4.82	5.03	5.20	5.35	5.49	5.61	5.71	5.81	5.90	5.98	6.06	6.13	6.20	6.27	6.33
12	3.08	3.77	4.20	4.51	4.75	4.95	5.12	5.27	5.39	5.51	5.61	5.71	5.80	5.88	5.95	6.02	6.09	6.15	6.21
13	3.06	3.73	4.15	4.45	4.69	4.88	5.05	5.19	5.32	5.43	5.53	5.63	5.71	5.79	5.86	5.93	5.99	6.05	6.11
14	3.03	3.70	4.11	4.41	4.64	4.83	4.99	5.13	5.25	5.36	5.46	5.55	5.64	5.71	5.79	5.85	5.91	5.97	6.03
15	3.01	3.67	4.08	4.37	4.59	4.78	4.94	5.08	5.20	5.31	5.40	5.49	5.57	5.65	5.72	5.78	5.85	5.90	5.96
16	3.00	3.65	4.05	4.33	4.56	4.74	4.90	5.03	5.15	5.26	5.35	5.44	5.52	5.59	5.66	5.73	5.79	5.84	5.90
17	2.98	3.63	4.02	4.30	4.52	4.70	4.86	4.99	5.11	5.21	5.31	5.39	5.47	5.54	5.61	5.67	5.73	5.79	5.84
18	2.97	3.61	4.00	4.28	4.49	4.67	4.82	4.96	5.07	5.17	5.27	5.35	5.43	5.50	5.57	5.63	5.69	5.74	5.79
19	2.96	3.59	3.98	4.25	4.47	4.65	4.79	4.92	5.04	5.14	5.23	5.31	5.39	5.46	5.53	5.59	5.65	5.70	5.75
20	2.95	3.58	3.96	4.23	4.45	4.62	4.77	4.90	5.01	5.11	5.20	5.28	5.36	5.43	5.49	5.55	5.61	5.66	5.71
24	2.92	3.53	3.90	4.17	4.37	4.54	4.68	4.81	4.92	5.01	5.10	5.18	5.25	5.32	5.38	5.44	5.49	5.55	5.59
30	2.89	3.49	3.85	4.10	4.30	4.46	4.60	4.72	4.82	4.92	5.00	5.08	5.15	5.21	5.27	5.33	5.38	5.43	5.47
40	2.86	3.44	3.79	4.04	4.23	4.39	4.52	4.63	4.73	4.82	4.90	4.98	5.04	5.11	5.16	5.22	5.27	5.31	5.36
60	2.83	3.40	3.74	3.98	4.16	4.31	4.44	4.55	4.65	4.73	4.81	4.88	4.94	5.00	5.06	5.11	5.15	5.20	5.24
120	2.80	3.36	3.68	3.92	4.10	4.24	4.36	4.47	4.56	4.64	4.71	4.78	4.84	4.90	4.95	5.00	5.04	5.09	5.13
x	2.77	3.31	3.63	3.86	4.03	4.17	4.29	4.39	4.47	4.55	4.62	4.68	4.74	4.80	4.85	4.89	4.93	4.97	5.01

Appendix 11

The following symbols are the ones we have adopted for use. Whilst most of them are in general use, variations are to be found in statistical literature.

$<$	less than: $3 < 4$
$>$	more than: $2 > 1$
x	the numerical value of an observation: eg wing length x mm; also the value of a frequency class
f	frequency; the number of observations in a frequency class x
y	the numerical value of an observation of a second variable from the same sampling unit from which x is taken: e.g. wing length x mm tail length y mm
x^2	x squared
$\sqrt{x}$	square root of x
N	number of sampling units in a population
n	number of sampling units (or observations) in a sample
n_i	the 'i th' observation in a series of n observations
Σ	capital sigma: the sum of
Σx	the sum of all values of x in a series of n observations
$(\Sigma x)^2$	the square of the sum of x in a series of n observations
Σx^2	the sum ot the squares of x in a series of n observations
μ	mu: population mean
$\bar{x}$	x bar: sample mean of n values of x
$\bar{x}'$	x bar primed: derived mean obtained from transformed values of x
$\bar{y}$	y bar: sample mean of n values of y
σ	sigma: population standard deviation
s	estimate of σ from sample data
σ^2	sigma squared: population variance
s^2	estimate of σ^2 from sample data
p	probability
z	standard deviation unit of the normal curve; test statistic in the z test
ν	nu: degrees of freedom
H_0	Null Hypothesis
H_1	alternative to a Null Hypothesis
χ^2	chi square: test statistic of the chi square test
r	product moment correlation coefficient of a sample
ρ	rho: product moment correlation coefficient of a population
r^2	coefficient of determination
r_s	Spearman rank correlation coefficient
a	intercept of a regression line on the y-axis; number of samples being compared
b	gradient of a regression line (also known as the regression coefficient)
t	test statistic of the t test
U	test statistic of the Mann-Whitney U-test
T	Test statistic of the Wilcoxon's test for matched pairs
F	test statistic of the F test
q	test statistic of the Tukey test
G	test statistic in the G -test

Bibliography and Further Reading

Bibby, C.J., Burgess, N.D., & Hill, D.A. (1992). *Bird Census Techniques*. Academic Press, London.

Essential reading for ornithologists planning a research programme. It explains how to collect reliable data.

Bishop, O.N. (1983) *Statistics for Biology*. Longmans.

An introduction to statistics for biologists which gives useful support to this Guide. A section on planning experiments is helpful.

Cohen, L. & Holliday, M. (1982). *Statistics for Social Scientists*. Harper & Row.

Contains a wider array of non-parametric tests, and includes programs in BASIC for most of the tests described in this Guide.

Elliott, J.M. (1969) *Some Methods for the Statistical Analysis of Benthic Invertebrates*. Freshwater Biological Assn. Pub. No. 25.

Although written for the freshwater biologist, this little book gives an excellent account of treatment of count data.

Fowler, J.A & Cohen, L.(1990) *Practical Statistics for Field Biology*. John Wiley.

An introductory text in similar format and scope to this Guide, with further treatments of probability distribution, frequency distributions and extended models of Analysis of Variance.

Moroney, M.J. (1953) *Facts from Figures*. Pelican.

Old-fashioned in style, this book makes interesting reading and presents different angles on problems.

Pearson, E.S. & Hartley, H.O. (1962) *Biometrika Tables for Statisticians*. Vol. 1. Cambridge University Press.

A reference work of statistical tables.

Sokal, R.R. & Rohlf, F. (1969) *Biometry* Second Edition. Freeman.

A standard text for biologists who wish to extend their knowledge of statistics beyond that dealt with in this Guide.

Southwood, T.R.E. (1978) *Ecological Methods*. Second Edition. Methuen.

A standard textbook on ecological methods. Includes a comprehensive section on mark/ recapture methods which is of interest to ringers.

INDEX

accuracy, 19
ANOVA - see Analysis of Variance
analysis of variance, meaning of, 116
    interaction in,
    one-way tests, 118
    in regression, 113
    with single observations, 129
    two-way tests, 122
arcsine transformation, 52
arithmetic mean, 29
association, tests for, 68,70
average, 29
    comparing, 104
axis, *x* and *y*, 87

bar graph, 22
bias, 12
bimodal distribution, 32
bivariate data, 20

Central Limit Theorem, 55
Chi-square tests, 68
class interval, 20
coefficient, 17
    correlation, 81
    of determination, 84,99
    regression, 88
    of variation, 37
common bird census, 9, 70
confidence interval of a mean, 55-57
    of a proportion, 61
confidence zone of a regression line, 94
constant effort site, 85
contingency tables, 70, 75-77
count data, 11, 57
counting things, 11-12
correlation, the meaning of, 80
    coefficient, 81
    significance of, 81
    strength of, 81
critical probability, 41
curved relationships, 98

data, definition of, 11
    bivariate, 20, 26
    continuous, 16, 23
    count, 57

discontinuous, 16
distribution-free, 47
normal, 47
presenting, 21-28
processing, 15
transformation of, 49-53
degrees of freedom, 35, 38, 70
dependent variable, 88
derived variables, 17-18
distribution, normal, 42
    frequency, 20
    t-, 46
Dodo, 39
dot diagram, 21

error, sampling, 54
    of a regression line, 93
    Type 1 and Type 2, 66
estimates, 54-61
    in regression, 96

F test, 111
frequency, class, 20
    aggregated, 24
    analysing, 68
    curve, 24
    distribution, 20, 23, 31
    expected, 41
    observed, 69
    polygon, 24
    proportional, 39
    table, 19

G tests, 74-78
goodness of fit, 68
gradients, in regression, 87
graph, bar, 22
    pie or circle, 27
    kite, 25

histogram, 22-23
homogeneity, test for, 68, 74
    of variance, 118
hypothesis, experimental and statistical, 62
    null, 63
independence, 13
    tests for, 68, 71
independent variables, 88, 97
interaction, in *ANOVA*, 122